Understanding Telecommunications and Lightwave Systems

IEEE PRESS Understanding Science & Technology Series

The IEEE PRESS Understanding Science & Technology Series treats important topics in science and technology in a simple and easy-to-understand manner. Designed expressly for the nonspecialist engineer, scientist, or technician as well as the technologically curious—Each volume stresses practical information over mathematical theorems and complicated derivations.

Books in the Series

Deutsch, S., *Understanding the Nervous System: An Engineering Perspective*

Evans, B., *Understanding Digital TV: The Route to HDTV*

Hecht, J., Sr., *Understanding Lasers: An Entry-Level Guide*, Second Edition

Kamm, L., *Understanding Electro-Mechanical Engineering: An Introduction to Mechatronics*

Kartalopoulos, S. V., *Understanding Neural Systems and Fuzzy Logic: Basic Concepts and Applications*

Nellist, J. G., *Understanding Telecommunications and Lightwave Systems: An Entry-Level Guide*, Second Edition

Sigfried, S., *Understanding Object-Oriented Software Engineering*

Ideas for future topics and authorship inquiries are welcome. Please write to the IEEE PRESS Understanding Science & Technology Series.

Understanding Telecommunications and Lightwave Systems

An Entry-Level Guide

SECOND EDITION

John G. Nellist

Consultant, Sarita Enterprises Ltd.

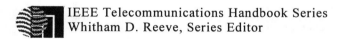

IEEE Telecommunications Handbook Series
Whitham D. Reeve, Series Editor

IEEE
PRESS

The Institute of Electrical and Electronics Engineers, Inc., New York

This book may be purchased at a discount from the publisher when ordered in bulk quantities. For more information contact:

IEEE PRESS Marketing
Attn: Special Sales
445 Hoes Lane
P.O. Box 1331
Piscataway, NJ 08855-1331
Fax (908) 981-9334

Printed in the United States of America
10   9   8   7   6   5   4   3   2

ISBN 0-7803-1113-2
IEEE Order Number: PP4665

Library of Congress Cataloging-in-Publication Data

Nellist, John G.
    Understanding telecommunications and lightwave systems / John G.
Nellist. — 2nd ed.
        p.   cm.
    Includes bibliographical references and index.
    ISBN 0-7803-1113-2 (paper)
    1. Telecommunication. 2. Optical communications. 3. Fiber
optics. I. Title.
TK5102.7.N45 1995
621.382—dc20   95-11535
                                                    CIP

Contents

Understanding Telecommunications and Lightwave Systems

Introduction

There is a growing recognition throughout the world that future economic and social progress will depend on effective communications. The common goal is a worldwide digital network over which interactions among various terminals, computers, and data bases can take place as universally as people talk to people today.

Communications and information processing have now become major growth industries. The speed and force with which this technology is moving is not recognized yet by the general public, but the influence it will have on how people live and work will be global and unalterable.

Lightwave communication systems have clearly emerged as the communications system of the future. This technology is now the most cost-effective method for the rapid transmission of digital information.

All over the world, optical fibers are replacing other transmission media in a wide variety of applications, including long-haul trunking systems, metropolitan telephone networks, undersea transmission links, and the local loop to the home. Optical fiber provides the important requirements of wide bandwidth, reliability and security. It is immune to most of the technical and regulatory constraints of traditional communication systems.

Understanding Telecommunications and Lightwave Systems is written in layman's language using a minimum of technical terminology. The primary objectives of this book are to give the entry-level student or the nontechnical manager an appreciation of this technology and to explain the basic principles that apply to the design of a telecommunications system.

Sections 2 through 8 provide an overview of the basic principles and technical standards of the present telephone network. Building on this base, sections 9 through 16 examine the current technologies of microwave radio, satellite communications, digital switching, video transmission and cellular phones. However, the main focus of this book is on lightwave systems, which are covered in section 17. This section provides an overview of this emerging technology and its impact on long distance networks, metropolitan telephone systems and the local loop to the business office and the home. Section 18 covers local area networks (LANs) and the integrated services digital network (ISDN). Section 19 takes a look at the future broadband integrated services digital network (BISDN).

Although this book deals primarily with the technical and regulatory trends in North America, the basic technology is common to telephone networks throughout the world. They differ only in the political, regulatory and societal conditions existing in each country.

We hope this book will provide an overview of telecommunications as well as an introduction to the exciting technology of lightwave communications. With the range of applications continuing to expand with growing speed, this technology promises to turn the electronic age into the age of optics.

1 The Evolution of Telecommunications

For centuries, long distance communications had been carried out by means of signal fires, lamps and flashing mirrors. However, the electric telegraph developed by Samuel Morse around 1835 actually launched long distance communications. By 1843, a telegraph line was constructed from Washington, D.C., to Baltimore. The first message sent over this line was "What hath God wrought?" which was certainly a good question in view of the developments which followed.

By 1850, Western Union was formed in Rochester, New York. Its purpose was to carry messages in a coded form (a dot and a dash) from one person to another over a privately controlled but publicly accessible network.

Alexander Graham Bell demonstrated his invention, the telephone, on March 10, 1876. Bell displayed the telephone at the 1876 World's Fair in Philadelphia. The Bell Telephone Company was formed in 1877 to commercially produce the telephone. In the same year, Western Union created the American Speaking Telephone Company as a competitor to Bell. By 1880, the American Bell Telephone Company was organized to serve as the parent company. Two years later, Western Electric Company was purchased to ensure a ready supply of telephones and related equipment. In 1885, the company was incorporated and became the American Telephone and Telegraph Company (AT&T).

Guglielmo Marconi, an Italian physicist, was the inventor of wireless communications, radio telegraphy. At the turn of the century many scientists believed that the curvature of the earth would limit practical use of the wireless to a distance of 100–200 miles. However, in 1901, Marconi proved them wrong by transmitting a message across the Atlantic Ocean.

In 1906, Lee DeForest announced his invention, the audion tube, the forerunner of the electron tube. The electron tube was used in radio to pick up faint electromagnetic signals and boost them a thousand times stronger than the received signal. Although it was intended to improve the sensitivity of radiotelegraph receivers, telephone engineers saw its potential for boosting long distance telephone signals as well.

By 1910, AT&T had gained control of Western Union. The U.S. Department of Justice threatened an antitrust suit against AT&T in 1913, so they agreed to dispose of Western Union. In 1934, the U.S. Congress created the Federal Communications Commission (FCC) and defined its powers in the Communication Act (1934). The FCC has jurisdiction over interstate and foreign commerce in communications but not telecommunications within a state. This is regulated by the state public utility commissions.

In Canada, the Trans-Canada telephone system was turned up on January 25, 1932, using an all-Canadian route. The system consisted of two open wire pairs strung on poles with voice repeaters spaced about 200 miles apart. In 1948, an additional open wire pair was strung across Canada to permit the installation of the first three-channel carrier system. This is a method of carrying more than one telephone circuit over a pair of open wires at the same time. Ten years later, the first microwave radio (TD-2) was turned up for service across Canada. The microwave radio provided better quality long distance circuits and more reliability than the open wire.

Scientists at Bell Laboratories in the United States introduced the transistor in 1947. The transistor is a solid-state device and does not require a heated cathode like the vacuum tube. Within a few years, the Bell Telephone System contained millions of transistorized elements.

By 1959, Texas Instruments and Fairchild Semiconductor successfully produced integrated circuits with transistors, capacitors and resistors placed on a square of silicon. Now, with an entire set of integrated circuits mounted together on a board, the whole board, costing only a few dollars, could be removed and replaced in the event of a problem.

In 1962, Bell Labs designed the first commercial pulse code modulation (PCM) cable carrier system. The introduction of the Bell System's T-carriers was the beginning of a trend toward digital communication in the telephone network.

Up until 1968, AT&T specified that only equipment furnished by AT&T could be attached to AT&T facilities. The Carter Electronic Corp. wanted to connect their mobile radio system to the telephone network. In the landmark 1968 Carterfone Decision, the FCC ruled that the AT&T restriction was unreasonable. This ruling gave birth to the interconnect industry, speeding the development of private switchboards and other devices for interconnection to the telephone network.

In 1969, a second landmark decision by the FCC permitted Microwave Communications, Inc. (MCI) to begin construction of a microwave radio system from St. Louis to Chicago. This private line system offered direct competition with Bell Telephone's toll network. This decision opened the door to other specialized common carriers.

The world's first commercial communication satellite was launched on April 6, 1965, by the Communications Satellite Corporation (Comsat) in the United States. This satellite, named *Intelsat 1* or *Early Bird,* was placed in geostationary equatorial orbit over the Atlantic Ocean. Although the United States was a pioneer in the establishment of an international satellite network, Canada had the first domestic satellite system. The satellite, designated *Anik A1,* was launched in November 1972 and began commercial service on January 11, 1973.

5

The development of optical fiber for use in the telecommunications industry did not begin until the mid 1960s. It was initiated in 1966 by the publication of a paper by Dr. C.K. Kao (ITT England), in which he stated that pure optical fiber was theoretically capable of guiding a light signal with very little loss. By 1970, Corning Glass Works in the United States was able to produce a fiber of sufficient purity for use in telecommunications. To make the fiber, Corning used a method of synthesizing silica glass. The raw materials were vaporized and deposited inside a length of quartz glass tubing, which was then collapsed into a rod and drawn into a fiber. The technology for the other two key elements of a fiber optic system, the laser and photodiode, had been developed independently during the previous ten years.

Up until 1982, AT&T provided 85% of all local telephone service and nearly 97% of all long distance telephone service. Then in 1982, AT&T and the Justice Department agreed to an antitrust settlement worked out by Judge Harold H.Greene. It required the divestiture of the 22 local operating companies. The United States was divided into 160 local access and transport areas (LATAs). The Bell operating companies (BOCs) were allowed to provide only local telephone service within the LATAs. They were forbidden from providing inter-LATA service. Each of the 22 companies was incorporated into one of seven regional Bell Operating Companies (RBOCs) as shown in Figure 1-1.

On January 1, 1984, AT&T's monopoly of telephone communications in the United States was over. The seven RBOCs are now all independent of AT&T, but they are restricted to providing local telephone service within their own area. However, subject to Judge Greene's approval, they are entering many new business areas.

The deregulation of the long distance telephone industry in the United States created opportunities for other telephone carriers to establish networks of their own. AT&T's major competitors, MCI and Sprint, have built nationwide lightwave (fiber optic) networks similar to the AT&T network.

Figure 1-1   **Seven Regional Bell Operating Companies**

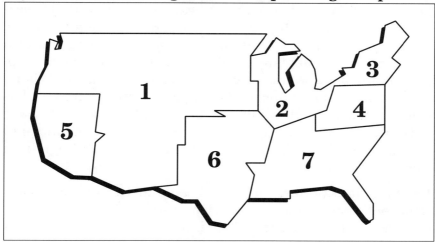

1 US West

Mountain Bell
Northwestern Bell
Pacific Northwest Bell

2 Ameritech

Illinois Bell
Indiana Bell
Michigan Bell
Ohio Bell
Wisconsin Bell

3 Nynex

New England Telephone
New York Telephone

4 Bell Atlantic

Bell of Pennsylvania
Diamond State Tel
New Jersey Bell
The Chesapeake & Potomac Cos

5 Pacific Telesis

Pacific Bell
Nevada Bell

6 Southwestern Bell Corporation   Southwestern Bell

7 Bell South

South Central Bell
Southern Bell

7

In Canada, the deregulation of the telecommunications market is underway, and new common carriers have started to surface. The largest competitor for the profitable long distance toll business is Unitel Communications. AT&T has acquired a stake in Unitel and plans to take an active role in the development of its toll network. Stentor, which is an alliance of Canada's major telecommunications companies (see Figure 1-2), has constructed a 7,000 kilometer transcontinental lightwave network. Stentor has a five-year, $300-million long distance phone technology sharing agreement with MCI, signed in December 1992.

In the United States, the Clinton administration has shown great enthusiasm for the creation of an "Information Superhighway" across America. This issue has captured the interest of telephone companies and cable television companies throughout North America. This national information infrastructure (NII), as it is called by the White House task force, will be an invisible, seamless, dynamic web of networks and information resources that will simultaneously carry limitless amounts of information in a variety of formats including voice, video, text and multimedia to unlimited locations. As a result, Tele-Communications Inc. (TCI), one of the largest cable television companies, has begun to build a vast cross-country lightwave network capable of delivering 500 channels of entertainment and electronic information to its subscribers. In Orlando, Florida, Time Warner, the largest cable television company, has created an interactive home entertainment and communication network for 4,000 subscribers. They will be able to call up a wide range of movies on demand, interactive video games and home shopping. Eventually, Time Warner wants to offer personal communications services (PCS), a mobile, wireless telephone system, with direct access to long distance telecommunications companies.

These changes create a significant threat to the RBOCs or "Baby Bells" that dominate the local telephone market. In addition, McCaw Cellular Communication Inc. has sold 33% of its

Figure 1-2   **Stentor**

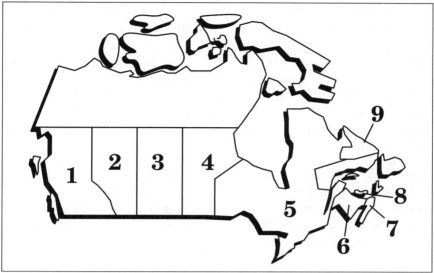

1   British Columbia Telephone Company
2   Alberta Government Telephones
3   Saskatchewan Telecommunications
4   Manitoba Telephone System
5   Bell Canada
6   New Brunswick Telephone Company
7   Maritime Telegraph and Telephone Company
8   Island Telephone Company
9   Newfoundland Telephone Company
Telesat Canada (satellite communications throughout Canada)

equity to AT&T. This gives AT&T its first foothold back in the local market and the means of bypassing the Baby Bells' access charges. The most important issue facing the RBOCs is which industry will dominate the provision of new multimedia services to the American home. As a result, US West has bought a 25% stake in Time Warner. Pacific Bell is spending $16 billion to create a multimedia communications superhighway across California. Nynex has invested $1.2 billion in Viacom, a cable television firm. Viacom recently acquired control of Paramount Communications Inc., owner of the last big independent film and television production company. Bell South plans to invest $1.5 billion in the QVC network, a home shopping firm. Television network CBS Inc. and QVC have recently announced a multibillion dollar merger. However, the biggest takeover was Bell Atlantic's plan to buy TCI for $16 billion, but the deal collapsed when the companies could not agree on a final offer.

Meanwhile, MCI, the second largest long distance company, is planning to spend $20 billion to build a transcontinental information superhighway. This will include the building of local telephone networks in 20 major US cities, and will put MCI in direct competition with the Baby Bells for the first time.

In 1996, SWB Company (the Southwest Bell Company RBOC) purchased Pacific Telesis (another RBOC) while NYNEX and Bell Atlantic (two other RBOCs) merged, putting back together some of what Judge Greene rent asunder in his "Modified Final Judgement" of 1984.

This is an example of how technological change is blurring the distinction between telecommunications, video transmission and computer industries, due to the ability to convert audio, video and data into digital form and transmit it over lightwave networks.

This new marketplace of the twenty-first century will force the merger of telecommunications, television, computers, consumer electronics, publishing and information services into a single interactive information industry. This will enable us to tap into and expand our vast resources of creativity and knowledge and lead to the development of products, services and industries that are beyond our imagination today. The revenue generated by this mega industry is estimated to reach $3.5 trillion worldwide by the year 2001.

2 Analog Transmission

When a person speaks, the vocal cords vibrate, producing sounds which are carried to the mouth. The sounds produced in speech contain frequencies which are in the 100 to 10,000 hertz (Hz) frequency band.

The notes produced by musical instruments occupy a much wider frequency band than that occupied by speech. Some instruments have a fundamental frequency of 50 Hz or less, while many other instruments can produce notes in excess of 15,000 Hz.

When sound waves enter the ear, they cause the eardrum to vibrate, producing signals. These signals, in the form of electric currents, are sent to the brain where they are interpreted as sound.

The human ear can distinguish frequencies between 30 Hz and 16,500 Hz. The average human voice ranges between 200 Hz and 5,000 Hz, as shown in Figure 2-1. Telephone company circuits operate over a range of frequencies from 300 Hz to 3,400 Hz. This is sufficient to make a person's voice recognizable and understandable.

Sound moves through the air in waves. The shorter the wavelength, the higher the frequency, or pitch, of the sound. Most sound waves, including our voices, are made up of many different frequencies and degrees of loudness.

Electricity moves through telephone wires in much the same way as sound waves move through the air. In the case of electricity, electrons bump against other electrons, sending their energy from one end of the wire to the other. Speech is transmitted by these electrical waves with the electricity in the wires vibrating in the same pattern as the sound waves.

Referring to Figure 2-2, when a person speaks into the mouthpiece of a telephone, the sound waves made by the vibration of the vocal cords strike a thin diaphragm, causing it to vibrate. As the sound waves compress the air against the diaphragm,

Figure 2-1 The Spectrum of Human Voice

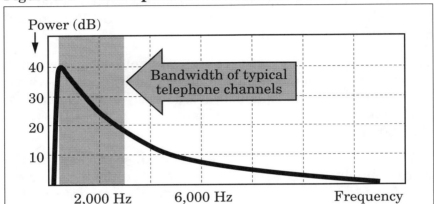

Figure 2-2 Analog Transmission

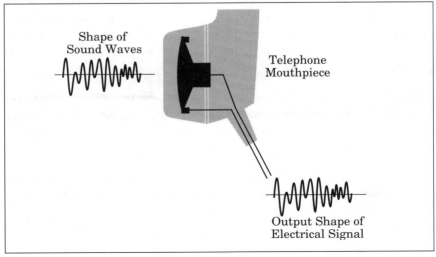

tiny carbon granules through which electric current is flowing are packed closely together. This creates a good electrical path for the current.

When the sound waves become less dense, the diaphragm springs back to its original position and the granules of the carbon move farther apart. This change reduces the flow of current. The diaphragm moves back and forth many hundreds of times per second in a pattern corresponding to the sound waves striking it. Thus, the amount of electricity flowing through the carbon granules varies, generating a signal. It is this electrical signal that is sent through the wires to the receiver at the other end.

At the receiver end, the electrical signals drive an electromagnet, whose varying force causes a diaphragm in the receiver to vibrate. The vibrations passing through the air between the receiver diaphragm and your ear are duplicates of the sound waves that struck the diaphragm of the transmitter at the other end.

The direct current (DC) for the telephone instrument is provided by a 48-volt battery at the local telephone office and fed to the telephone over a twisted pair of copper wires designated "Tip" and "Ring." When the user lifts the handset from the cradle, the instrument draws the direct current from the line when the switch hook closes (see Figure 2-3). The carbon microphones in the original handset have been replaced by miniature self-polarized capacitive microphones, called electrets. The original electromagnetic receivers have been replaced by piezoelectric receivers.

In the late 1960s and early 1970s, a transducer for an electronic ringer was introduced that eventually replaced the hammer and bell ringer that was invented in 1878. The rotary dial has been largely replaced by the push-button keypad. The keypad controls the circuitry that generates either dial pulses or the tones for touch-tone dialing.

Figure 2-3   **Telephone Power Supply**

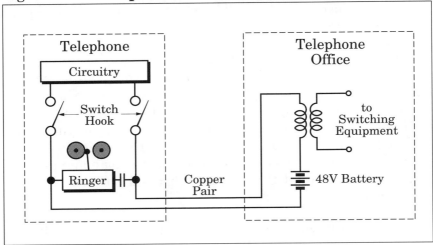

Conclusion

Information is produced and transmitted over the telephone network as electrical signals. These signals have two forms: analog and digital. Analog signals are continuous and can be thought of as electrical voltages that vary continuously with time. Digital signals are a series of on/off pulses and will be examined in Section 3.

Transmission in the telephone network was completely analog until 1962, when digital transmission was introduced. Some long distance lines and most customer loops are still analog because of the tremendous investment that telephone companies have in this equipment.

However, due to the new competition in the telephone industry and the introduction of new technologies, this situation is changing rapidly, as we will see in the following sections.

Review Questions for Section 2

1. What is the frequency range of the human ear?

2. What is the frequency range of the average human voice?

3. What is the frequency range of a typical telephone company circuit?

4. Explain how the telephone set is provided with power from the telephone office.

5. Identify the device that replaced the original carbon microphones in the telephone set.

6. Identify two types of signaling that may be accomplished with a telephone set.

3 Digital Transmission

Digital transmission means that voice, data and video signals can be sent in digital form as a stream of on/off pulses. Computer data can be transmitted directly over a digital transmission channel. Analog information, such as the human voice, needs to be converted to digital form to be transmitted. This conversion is performed by a device called a **codec** (from **co**der / **dec**oder).

Any analog signal, hi-fi music, video or photographs, can be digitized into a bit stream and transmitted over a digital channel. The data transmission rate (bits per second) required is dependent on the range of frequencies of the analog signal as well as on the number of different amplitude levels to be reproduced.

A telephone call, when digitized, uses 64 kb/s (kilobits per second) to faithfully reconstruct the original analog waveform. Sampling must be done at a rate of at least twice the highest frequency in the wave. Telephone companies have established 4 kilohertz (kHz) as the standard "bandwidth" of a voice channel. Therefore, sampling is done at 8 kHz, and because an 8-bit word is assigned to each sample, the data rate is 8 x 8 or 64 kb/s. Television and hi-fi music require a much higher transmission rate. A detailed description of this process will be found in Section 4.

A transmission medium such as copper wire pairs, coaxial cable, microwave radio or lightwave can be designed to carry information in either analog or digital form (see Figure 3-1).

16

Figure 3-1   **Analog Signal and Digital Signal**

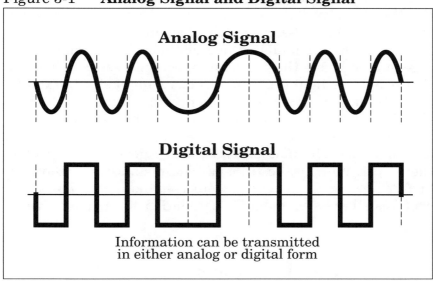

Analog Signal

Digital Signal

Information can be transmitted
in either analog or digital form

Digital transmission conveys information in the form of symbols called bits (binary digit). A bit is always one of two things
(pulse or no pulse; mark or space; 1 or 0; on or off) and can
be nothing else. By contrast, analog transmission conveys
information by waves of continuously varying amplitude or
frequency, and every minor impairment of the wave implies a
corresponding degradation of the information content. In
digital transmission, however, minor variations of the signal
do not change the information content. A pulse is still a pulse
even if it is distorted, and only a major distortion can change a
pulse to no pulse, or no pulse to an invalid no pulse.

In any transmission system, the received signal differs from
the transmitted signal because of noise, crosstalk and intermodulation. In analog transmission, the disturbance, once introduced, cannot be eliminated, and the information content is
degraded. In digital transmission, the degradation of the
received signal by the transmission system does not alter the
information content until the degradation becomes so severe

that the receiving equipment reads a pulse as no pulse. Below this threshold level, the information transfer is essentially perfect. Furthermore, because of this fact, digital transmission is not limited by accumulated degradation. The signal can be regenerated before the degradation becomes severe (shown in Figure 3-2). A digital repeater reads its input signal, extracts the information, and uses it to generate a brand new output signal for the next transmission section. The loss, noise, interference and distortion of the preceding transmission section are completely eliminated. Thus, with regenerative repeaters there is no limit to the transmission distance. By contrast, analog repeaters are not regenerative, and the output signal contains all the accumulated degradation of the input signal.

Figure 3-2   **Pulse Regeneration**

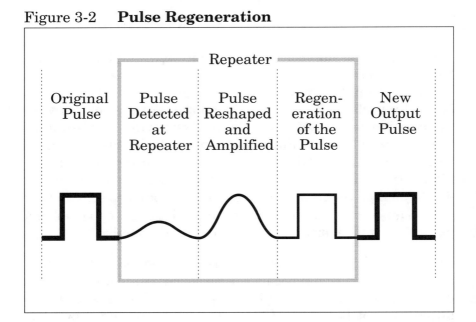

Conclusion

The telephone system is evolving from an analog network to a broadband integrated services digital network (BISDN) where the simultaneous transmission of voice, data and video will be possible. New digital switching machines, digital microwave systems and digital lightwave systems can all handle digitized voice and data. The picture is changing rapidly. Digital switches are extending their reach to digital PBXs (private branch exchanges) and digital telephones, giving business direct access to the digital network.

Codecs will continue to migrate from the switching machines and the PBXs down to residential telephones.

One day, the microphones and speakers in the telephone sets will be the only analog components remaining in the network.

Review Questions for Section 3

1. What is the device called that converts analog information into a digital format?

2. Are voice frequencies the only type of analog information which can be digitized and transmitted over a digital channel?

3. What bandwidth have telephone companies established for a standard voice circuit?

4. What is the data transmission rate for a standard voice circuit?

5. Describe the function of a binary digit (bit).

6. What is the major advantage of digital communications?

7. Describe the function of a regenerative repeater.

4 Basic Multiplexing Techniques

As telephone networks grew and traffic increased, additional circuits were needed to meet the growth. However, there was a physical limit to the number of wires that could be carried overhead on poles and crossarms or placed in underground ducts. It became obvious that more than one voice circuit must be carried over the same facility at the same time. As a result, a new technique was developed which was called carrier transmission. This technique allowed the original voice frequency (300 Hz to 3,400 Hz) to be shifted to a higher frequency by a process called multiplexing. This permits a number of voice channels to be transmitted over the same line. Multiplexing achieves this by the use of a bandwidth much larger than any of the individual channels, and currently uses one of two specific techniques: frequency division multiplexing (FDM) or time division multiplexing (TDM)

FDM uses a method of stacking the channels, with each channel occupying a different portion of the frequency spectrum, as shown in Figure 4-1. FDM has been the basic signal combining technique for analog carrier systems for the past 50 years.

To achieve frequency separation, each channel amplitude modulates a different carrier frequency. Twelve 4-kHz voice channels make up a group with a frequency range of 60 kHz to 108 kHz. Five groups can be combined to create a supergroup with a frequency range of 312 kHz to 552 kHz. However, most analog carrier systems have been replaced by digital carrier systems using TDM.

TDM uses a method of splitting time into narrow slices. Input signals are sampled one after the other at high speed, 8,000

Figure 4-1   **Frequency Division Multiplexing (FDM)**

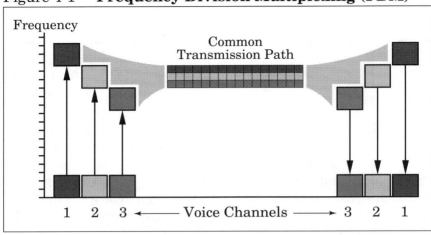

times per second. Only one sample of a specific signal occupies the channel at any particular time. Figure 4-2 shows a simplified arrangement with three analog signals being sampled and transmitted over one transmission path.

Pulse code modulation (PCM) combined with TDM is the most widely used method of transmitting analog signals over digital transmission facilities.

To facilitate digital transmission, it is necessary to convert the primary information into digital information for transmission and reconvert the digital information back to its primary form at the receiver. This is the function of the terminal equipment.

There are three steps to PCM: sampling, quantizing and encoding. Sampling must be done at a rate of at least twice the highest frequency being transmitted. Therefore, sampling is done at 8 kHz, and because an 8-bit word is assigned to each sample, the data rate is 8 x 8 or 64 kb/s, known as the DS-0 rate. Each voice channel is sampled 8,000 times per second, and the samples from 24 channels are interleaved (multiplexed) in time to form pulse amplitude modulation (PAM) frames as shown in Figure 4-3. Each frame consists of 24 amplitude

Figure 4-2   **Time Division Multiplexing (TDM)**

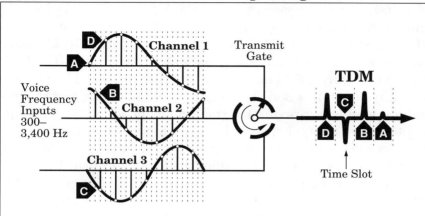

- Time division – a piece of each channel in turn by time
- Multiplex – all input channels share the common transmit path

Figure 4-3   **Pulse Amplitude Modulation (PAM)**

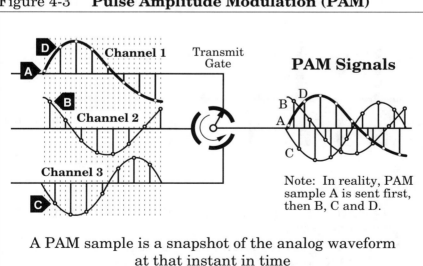

A PAM sample is a snapshot of the analog waveform
at that instant in time

samples, one from each channel, and frames occur at the rate of 8,000 times per second. The PAM frame is quantized as shown in Figure 4-4 by taking samples at specific intervals. The signal is then encoded by converting each amplitude sample into a code word consisting of eight binary digits (0 or 1) as shown in Figure 4-5. The eight bits of the code word can represent any one of $2^8 = 256$ distinct amplitude values, and the encoding process approximates the amplitude sample by assigning to it the nearest available code value. All eight bits are used for information encoding in five out of six frames. In every sixth frame, the eighth bit, called "the least significant bit" carries the signaling information. After each sequence of 24 8-bit words an additional framing bit is inserted to supply necessary synchronizing information for the receiving terminal. Thus each PCM frame consists of $(24 \times 8) + 1 = 193$ bits as shown in Figure 4-6. Because frames occur 8,000 times per second, the terminal bit rate is $193 \times 8,000 = 1,544,000$ bits per second.

Figure 4-4   **Quantizing PAM**

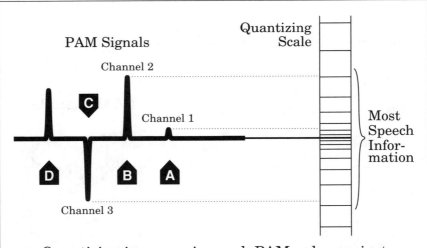

- Quantizing is measuring each PAM pulse against a reference scale
- Quantizing scale is nonlinear because most speech information is at a soft level

Figure 4-5   **Encoding to PCM**

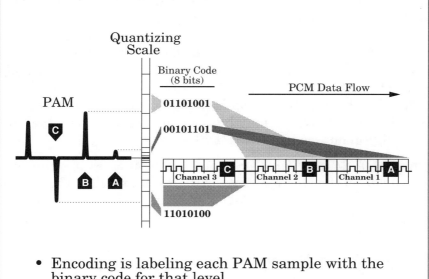

- Encoding is labeling each PAM sample with the binary code for that level
- The binary code is transmitted as digital pulses

Figure 4-6   **PCM Frame**

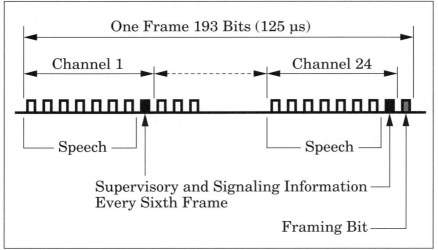

25

This bit stream (1.544 Mb/s) is transmitted between terminals over repeatered T-1 span lines which use one cable pair for each direction of transmission as shown in Figure 4-7. The regenerative repeaters are spaced approximately every 6,000 feet. As the bit stream passes through each repeater point, new clean pulses are reconstructed and regenerated. This process largely overcomes the problems of noise, interference, loss and distortion. For short distance applications, 9,000 to 12,000 feet, the 1.544 Mb/s bit stream may be transmitted using the high bit rate digital subscriber line (HDSL).

The DS-0 (64 kb/s) and DS-1 (1.544 Mb/s) rates are the first two levels of the North American digital hierarchy. The digital hierarchy consists of other levels and rates, which are explained in Section 6.

Figure 4-7   **PCM Carrier**

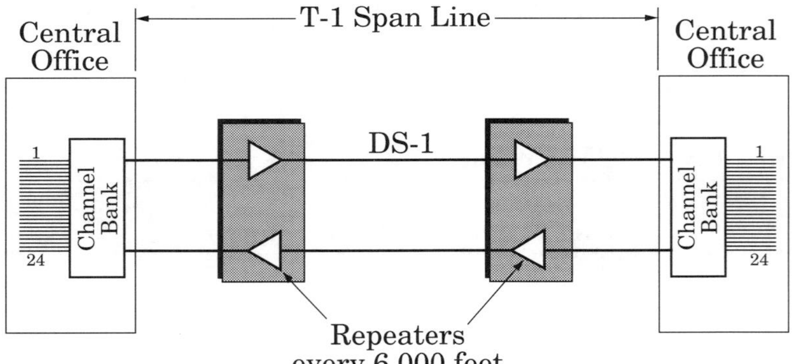

- Channel bank converts voice to digital code
- Multiplexes 24 conversations onto span line
- DS-1 digital bit rate of 1.544 Mb/s

Conclusion

Many voice circuits can be multiplexed together over a single transmission medium. There are two approaches to multiplexing: Analog – frequency division multiplexing (FDM), and Digital – time division multiplexing (TDM). In FDM, a number of voice circuits are combined, with each circuit given its own unique space in the frequency spectrum. TDM uses a method of splitting time into narrow slices. Combined with pulse code modulation (PCM), which encodes the signal into binary digits, it is the most widely used method of multiplexing analog signals over digital transmission facilities.

Review Questions for Section 4

1. Name two basic multiplexing techniques.

2. Why were carrier transmission and multiplexing techniques developed?

3. What is the most widely used method of multiplexing analog signals over digital transmission facilities?

4. Briefly describe the technique of time division multiplexing (TDM).

5. What are the three basic steps to pulse code modulation (PCM)?

6. In PCM, what is the sampling rate for a standard voice circuit?

7. How many bits are used to form each PCM frame?

8. What is the usual spacing between regenerative repeaters on a T-1 span?

5 Switching Hierarchy

Early telephone customers were linked by point-to-point lines as shown in Figure 5-1. This type of connection seems like a simple system. However, as the number of customers increases, the number of lines increases even faster. With over 165 million telephone customers in the United States and Canada, it would be physically impossible to connect all the telephones in this point-to-point manner.

The first telephone exchanges were manual and all the calls were established by operators. However, today most of the world uses automatic telephone switching systems employing either electromechanical relays or the latest computer-controlled electronic machines.

Figure 5-1   **Point-to-Point Connections**

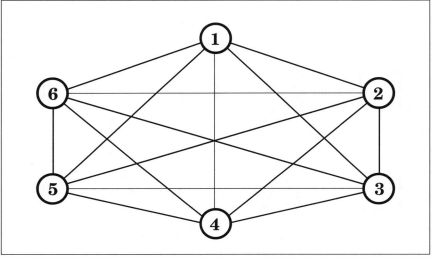

A central switching point permits the connection of customers in a simple star fashion as shown in Figure 5-2. All calls are routed through the central switching point, thus dramatically reducing the number of lines required.

As telephone usage grew, network switching evolved into a hierarchy that consisted of the five levels shown in Figure 5-3. Upon divestiture of the RBOCs from AT&T, the access by long distance companies to local networks changed and, for all prac-

Figure 5-2   **Central Switching Point**

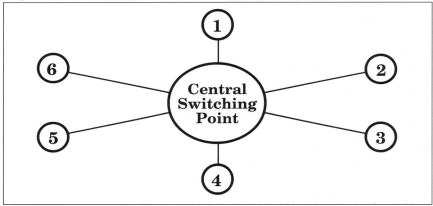

Figure 5-3   **Five-Level Switching Hierarchy**

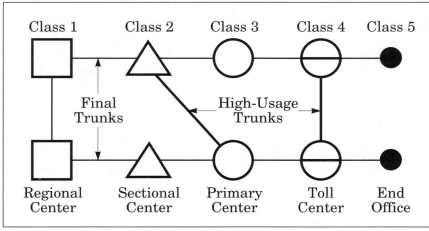

tical purposes, eliminated the traditional switching hierarchy in the United States. Just prior to divestiture at the lowest level there were some 20,000 Class 5 switching centers called **end offices**, which interface directly with the customer equipment via the "local loop." At the next level there were 1,300 or more **toll centers** (Class 4), so called because their usage implies higher rates. Then came 265 **primary centers** (Class 3) and 75 **sectional centers** (Class 2). At the top of the hierarchy were **regional centers** (Class 1), which numbered only 12 (10 in the United States and 2 in Canada). Traffic is always routed through the lowest available level in the hierarchy; if that level is busy, higher levels are selected.

It was not necessary that Class 4 or 3 offices always home on the next higher ranking office. Possible homing arrangements for each class of switching office are shown in the following table.

Rank	Class of Office	May Home On
End Office	5	Class 4, 3, 2 or 1
Toll Center	4	Class 3, 2 or 1
Primary Center	3	Class 2 or 1
Sectional Center	2	Class 1
Regional Center	1	All regional centers interconnected

In the United States, the divestiture has created the "access tandem." This is the gateway between the interexchange carrier's (for example AT&T, Sprint and MCI) point of presence (POP) and the exchange carrier's end office. End offices and access tandems may be served directly from any interexchange carrier location.

Routing rules between switching centers determine the selection of a fixed set of alternate routes using final or high-usage trunks. However, there are some limitations to the efficiency gains achievable with hierarchical alternate routing. A number of dynamic routing concepts are being used in the North American toll network. Dynamic nonhierarchical routing (DNHR) is AT&T's version of circuit-switched dynamic routing. DNHR is a centralized time-dependent routing scheme that increases network efficiency by taking advantage of the noncoincidence of busy hours in the North American toll network. Common channel interoffice signaling (CCIS) is used for signaling between telephone exchanges in the call path. CCIS is a system for exchanging signaling information between exchanges via a network of signaling links instead of on the individual voice circuits. Another routing scheme called dynamically controlled routing (DCR) is a centralized adaptive routing system developed by Northern Telecom. The DCR concept is an advanced and flexible form of dynamic routing that makes efficient use of the network resources.

Conclusion

Telephones are connected to each other via a hierarchy of switching centers. The lowest level switching center is called the end office. Customers can access other customers connected to other end offices toll free within the local exchange area.

All long distance calls are routed through the interexchange carrier switching offices. The hierarchical system which consisted of the toll center, sectional center and finally the regional center has been largely replaced by a less compartmentalized system in which dynamic routing concepts are applied on a network basis. The continuing introduction of new technology in the North American toll network will see a wider application of dynamic routing concepts in the future.

Review Questions for Section 5

1. List the five levels in the traditional switching hierarchy.

2. Does a Class 5 office always home on a Class 4 office?

3. Explain the term point of presence (POP).

4. Explain the difference between a high-usage trunk and a final trunk.

5. Identify two types of dynamic routing schemes.

6. Briefly describe the concept of common channel interoffice signaling (CCIS).

6 North American Digital Hierarchies

The design of the North American Digital Network has evolved around a series of hierarchical levels based on the DS-1 (1.544 Mb/s) primary rate. The DS-1 rate was established by the Bell Labs as the transmission rate for the first commercial pulse code modulation (PCM) cable carrier system back in 1962. This rate was chosen as an optimum rate for transmission over existing 6,000 foot spans of 22-gauge exchange grade cable. It has since been accepted as the basic building block for the North American digital hierarchy. The majority of transmission systems and multiplexers in use in North America today are electrically compatible at this rate, although signaling formats may vary.

The DS-1C (3.152 Mb/s) rate was derived by combining two DS-1 inputs and adding housekeeping pulses for frame alignment and synchronization.

The DS-2 (6.312 Mb/s) rate was derived by combining four DS-1 inputs and adding housekeeping pulses for frame alignment and synchronization.

The DS-3 (44.736 Mb/s) rate was derived by combining seven DS-2 inputs and adding pulses for frame alignment and synchronization, as shown in Figure 6-1.

The next step up in the digital hierarchy is the DS-4 (274.176 Mb/s) rate, which has traditionally been the highest level in the North American telecommunications hierarchy. However, all the new long-haul lightwave systems go beyond this rate due to the rapid development in this new technology. Their

Figure 6-1   **Digital Signal Hierarchy**

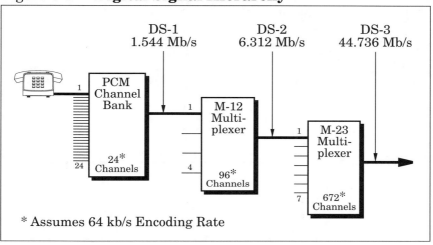

* Assumes 64 kb/s Encoding Rate

transmission rates are all multiples of DS-3 streams. The current rates are 565 Mb/s (12 DS-3s), 1.2 gigabits per second (Gb/s) (24DS-3s), and 2.4 Gb/s (48 DS-3s). Lightwave systems associated with the SONET use different rates, which are discussed in Section 17.

All of these digital bit streams use the same TDM techniques. For instance, the DS-3 bit stream is made up in an M-23 multiplexer by taking one bit at a time from each of the seven DS-2 inputs and sequentially interleaving them to form a single bit stream. Housekeeping bits are added at definite intervals carrying control information. The resultant DS-3 signal, along with five other DS-3 signals, can then be connected to an M-34 multiplexer, where the process is repeated again to create a DS-4 signal. The only digital cable carrier that used this bit rate (274.176 Mb/s) was the T4M system, which operated over coaxial cable and provided 4,032 voice circuits.

Various multiplexing schemes can be used to achieve conversion between the levels of the digital hierarchy. Figure 6-2 shows commonly accepted relationships as well as the names of the digital signal cross-connects used at each level.

Figure 6-2   **Digital Signal Building Blocks**

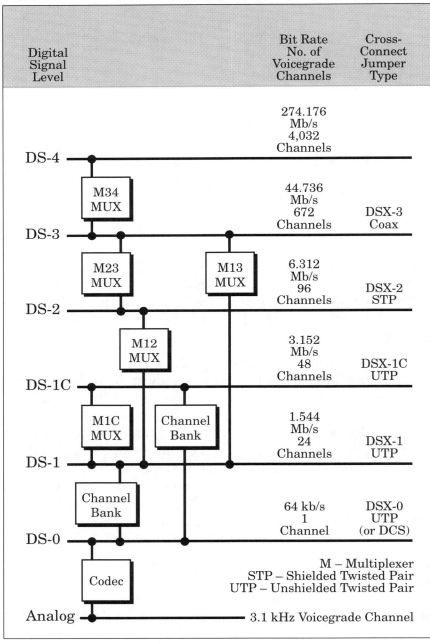

Digital Signal Level		Bit Rate No. of Voicegrade Channels	Cross-Connect Jumper Type
DS-4		274.176 Mb/s 4,032 Channels	
	M34 MUX	44.736 Mb/s 672 Channels	DSX-3 Coax
DS-3			
	M23 MUX M13 MUX	6.312 Mb/s 96 Channels	DSX-2 STP
DS-2			
	M12 MUX	3.152 Mb/s 48 Channels	DSX-1C UTP
DS-1C			
	M1C MUX Channel Bank	1.544 Mb/s 24 Channels	DSX-1 UTP
DS-1			
	Channel Bank	64 kb/s 1 Channel	DSX-0 UTP (or DCS)
DS-0			
	Codec		
Analog		3.1 kHz Voicegrade Channel	

M – Multiplexer
STP – Shielded Twisted Pair
UTP – Unshielded Twisted Pair

36

Conclusion

One of the first applications of digital signal processing (DSP) in the telephone network was the installation of pulse code modulation (PCM) carrier systems back in the 1960s. This was the introduction of PCM carrier systems and the acceptance of 64 kb/s as the standard for coding voice in the emerging digital network. While 64 kb/s PCM provides excellent transmission quality, it has been realized for some time that it is relatively inefficient in its use of transmission facility bandwidth. Essentially, there are two basic techniques which could be applied to increase the circuit capacity: digital signal interpolation (DSI) and low-bit-rate encoding (LBRE).

DSI is a process which time-shares a fixed number of digital voice channels between a larger number of talkers. Since the average speech activity in a two-way voice conversation is typically on the order of 30% to 40% in each direction, it is possible to fill in the speech gaps in one conversation with talk spurts from another conversation. Practical systems which provide a 2:1 concentration ratio (a doubling of transmission capacity) are commercially available. Higher concentration ratios are realizable when the network carries a high proportion of voice traffic. Lower concentration ratios are necessary with data traffic because data signals normally operate continuously on a given call.

LBRE is a technique which uses advanced signal-processing techniques to reduce the number of bits required to encode the voice signal. The most promising technique is to exploit the correlation between successive speech samples. The difference between the actual voice sample and an estimated (or predicted) value based on the immediate preceding voice samples is encoded and transmitted. This is referred to as adaptive differential pulse code modulation (ADPCM).

PCM carrier systems employing this technique with an effective encoding rate of 32 kb/s are used in many digital transmission systems. These LBRE systems initially were used to provide private line voice service but are now widely deployed throughout the public network. The 32 kb/s ADPCM has many advantages over DSI systems but does not provide as much potential efficiency and it does not provide good performance with voiceband data applications.

Also under development are 16 kb/s and 8 kb/s schemes. Subject to further studies, it is expected that these techniques will find additional applications within the toll network over the next few years because they have the potential to greatly increase the capacity of existing lightwave, digital radio and satellite systems.

Review Questions for Section 6

1. What are the bit rates for the first three levels in the North American digital hierarchy?

2. Explain the function of an M-23 multiplexer.

3. How many DS-3 bit streams can be carried on a 565 Mb/s carrier system?

4. Name two techniques which could be used to increase the circuit capacity of a PCM carrier system.

5. What does ADPCM stand for?

7 Transmission

The international telecommunications network is growing so rapidly that it is important to have a single set of consistent standards and objectives that apply worldwide. Although incompatibilities do exist between the European and North American networks, the degree of compatibility is remarkable. This is largely due to the International Telecommunication Union (ITU). This organization, headquartered in Switzerland, has over 140 member countries throughout the world. Its consultative committees carry out very detailed studies of world telecommunications and make recommendations for standardization.

In the United States, there are many standards-setting organizations which have been in existence for many years, such as the American National Standards Institute (ANSI) and the Institute of Electrical & Electronic Engineers (IEEE). Relatively new organizations, such as Committee T1 of the Alliance for Telecommunications Industry Solutions, Bellcore and the Telecommunications Industry Association (TIA), are concentrating on the standardization of network aspects and terminal equipment.

All of these organizations have done a creditable job in tracking the technology growth and producing coordinated standards.

The need for transmission standards to serve as a guide in the design, operation and maintenance of the telephone network has been recognized since the early days of the telephone. Over the years, the details of the approach have changed with the state of the art and the usage of the network, but the purpose has remained essentially unchanged: to provide a quality of service which is based on an appropriate balance between customer satisfaction and the price the customer must pay for the service.

The transmission level at any point in the telephone system is defined in terms of loss or gain in decibels referenced to 1 milliwatt (dBm) from an arbitrary reference point called the zero transmission level point (0 TLP). In other words, 0 dBm = 1 milliwatt, with a logarithmic relationship as the values increase or decrease. For example, in Figure 7-1, if a −10 dBm tone is sent at 0 dB TLP, then a level of −12 dBm would be measured at the −2 dB TLP. In an analog network, end offices are at 0 dB TLP and toll offices are at −2 dB TLP.

With the digital offices, the switched signals consist of bit streams with no meaningful levels. However, a digital test code sequence (DTS) is used for measurement purposes. This is a repetitive code sequence which, when converted to an analog signal, will produce 0 dBm at about 1,000 Hz.

Figure 7-1   **Transmission Level Point (TLP)**

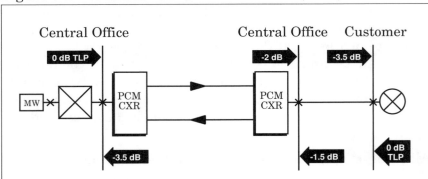

• An accessible point is designated 0 dB TLP

• All other level measurement points may now be referenced to the 0 dB TLP, i.e., if −10 dBm tone is sent at 0 dB TLP, then you would measure −12 dBm at the −2 dB TLP

40

The via net loss plan (VNL) for the North American switched network was introduced with direct distance dialing (DDD) in the late 1950s. The term DDD is used to describe long distance calls dialed by customers without assistance by operators. A VNL factor was developed for each of the several classifications of facilities (cable, carrier, radio). The classification was made on the basis of propagation time per unit distance.

The principle of the VNL plan was to increase loss as a function of propagation time, up to the point when an echo suppressor is considered necessary. The reasoning behind this was that the more an echo is delayed, the more annoying it is and hence the more it must be attenuated to reduce the annoyance. An echo suppressor was introduced at the point where the advantage of reducing echo level and the disadvantage of reducing voice level are subjectively comparable to reducing echo level by a suppressor.

Forty-five milliseconds (ms) was the objective for the maximum round-trip delay without echo suppression. The maximum loss on high-velocity intertoll trunks is about 2.9 dB, with current echo suppressor application rules calling for a suppressor at 3,000 kilometers (1,850 miles). Trunks longer than this are operated with an echo suppresser and at zero loss. (For a description of an echo suppressor and the more technologically advanced echo canceller refer to page 45.)

With the introduction of digital facilities and digital switching machines in the 1970s and 1980s, the VNL plan was no longer suitable for a mixed analog-digital network.

An all-digital network could be operated at zero loss, or even with gain to accommodate the characteristics of the 4-wire telephone set which would be used in this case. However, a more realistic scenario is a digital network with 2-wire customer loops (see Figure 7-2). This requires 4-wire to 2-wire conversion between the customer's local loop and the digital end office.

Figure 7-2   **Switched Digital Network (SDN)**

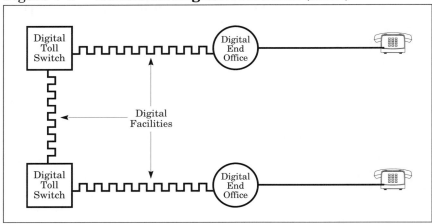

As a result, a new fixed loss plan (FLP) was introduced. The end office digital switch must be capable of inserting a loss of 0, 3, 5 or 6 dB, depending on the type of call and the trunks involved, into the receive leg of a connection.

Three types of trunks have been defined for the purpose of applying the fixed loss plan in the mixed network. These are illustrated in Figure 7-3 and defined as follows:

- Analog trunks are defined as trunks that use analog or mixed analog and digital facilities regardless of the type of switch they terminate in at each end, or are trunks which use digital facilities only but terminate in analog switches at both ends.

- Combination trunks are defined as trunks that use digital facilities only and terminate in a digital switch at one end and an analog switch at the other end.

- Digital trunks are defined as trunks that use digital facilities only and terminate in digital switches at both ends.

In the mixed network, either the VNL or the FLP is applied, based upon whether the trunk is analog, combination or digital.

Figure 7-3 **Mixed Analog-Digital Network**

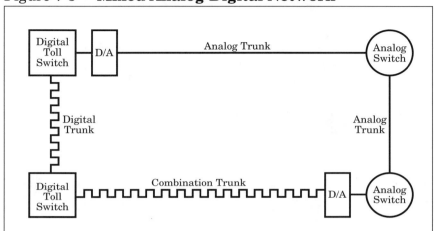

As more digital carrier systems, terminating on analog switches, were cut over to new digital switches, the channel banks (analog-to-digital (A/D) converters) were eliminated. This not only simplified the network from a maintenance standpoint but also greatly improved the overall transmission performance for the customer.

In an all-analog network, loss, noise and echo are significant transmission impairments which influence customer satisfaction (see Figure 7-4). In a digital network, digital trunks are, strictly speaking, noiseless and lossless, and most of the customer-to-customer loss is concentrated in the loops (see Figure 7-5).

Ultimately, in an all-digital network (with digital loops) loss and noise will no longer be significant parameters. However, an all-digital network requires built-in echo cancellation to reduce the effects of delayed echo on voice conversations and voiceband data transmission using 2-wire modems.

Figure 7-4   **Analog Network Connection**

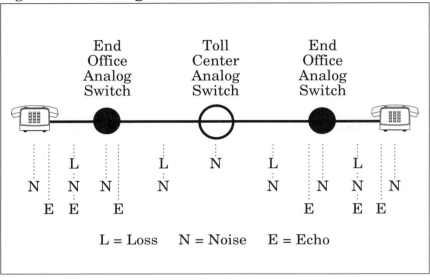

Figure 7-5   **Digital Network Connection**

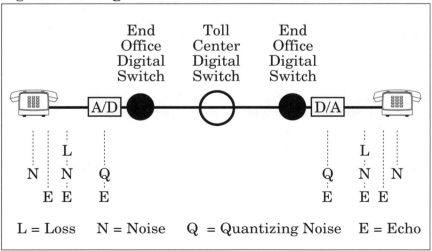

Echo Suppressors and Echo Cancellers

Echoes occur in a long distance telephone connection at any point where an impedance mismatch causes energy to be reflected. The point where most echoes occur in today's digital network is at the 4-wire to 2-wire conversion point in the end office. In analog end offices this is at the hybrid between the 4-wire interexchange circuit and the 2-wire switching system. In digital end offices, this is in the end office line circuit, which converts the 2-wire local loop to the 4-wire equivalent internal network in the digital switch. Two factors, loudness and delay, determine how objectionable an echo is. It is obvious that an echo that is inaudible will not be annoying, no matter how long it is delayed. However, if an echo is loud enough to be noticeable, it becomes more objectionable with increasing delay between it and the original signal. Given the long round-trip delay, echo can be a serious problem when a satellite link is in the transmission path.

On long paths, echo suppressors were originally used to control echoes. Figure 7-6 is a simplified diagram of a "split" echo suppressor which provides suppression for one direction of

Figure 7-6   **Split Echo Suppressor**

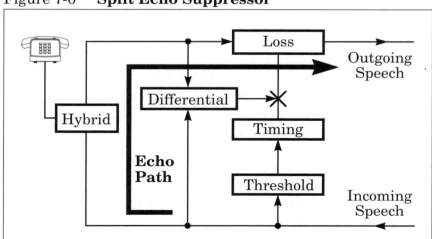

transmission at each end of the circuit. Basically, the echo suppressor is a pair of voice-actuated switches which, while one party is talking, inserts a high loss of 50 dB or more into the echo path. In Figure 7-6, incoming speech that is echoed in the hybrid will be attenuated on the outgoing speech path.

Most of the time, the parties at either end of the line speak in turn, so the suppressor is effective in removing the echo. However, interruptions and double-talking are part of the dynamics of conversation, that is, when both parties talk at the same time. To handle double-talk, a suppressor suspends echo suppression and changes to double-talk operation. The high loss is removed, and a small loss, approximately 6 dB, is inserted in the talker's receive path. This receive loss attenuates the speech as well as the echo. Also, the switching decisions require a finite amount of time and cannot be made with absolute accuracy. So some amount of speech clipping is inevitable, particularly on the first part of an interruption. Experience has shown that echoes during double-talk and speech clipping are more annoying when they are combined with satellite delay.

Another variation of the suppressor is the "full" echo suppressor, which provides suppression for both directions of transmission at one end only. This arrangement results in poorer break-in properties, so it is not recommended for satellite links and is limited to terrestrial circuits of less than 6,100 kilometers.

A more complex device called an echo canceller, using digital signal processing techniques, provides a much better solution to the problem and is the technology of choice in a modern digital network. As shown in Figure 7-7, the processor simulates the impulse response of the echo path and generates a replica of the actual echo signal. When one party is talking, the transmitted signal consists of speech only. When both parties speak simultaneously, the signal consists of both originating speech and echo from incoming speech. The echo replica is continuously subtracted from the transmit signal at the receiving end, thereby selectively cancelling the echo.

Figure 7-7   **Digital Echo Canceller**

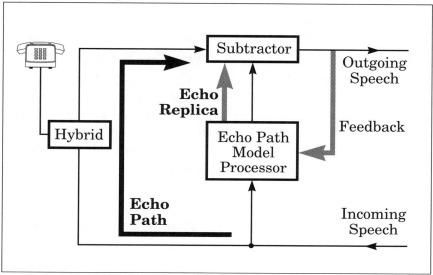

With the widespread introduction of domestic satellite systems into the network, there is a possibility of satellite links being connected in tandem. Such double-hop satellite circuits are undesirable because of the 1,080 ms delay (2 x 540 ms) (see Figure 10-4).

From a transmission point of view, in order to provide equivalent subjective performance on satellite circuits compared to terrestrial circuits, the objective is to equip all satellite trunks with echo cancellers. While canceller-equipped trunks in tandem provide acceptable performance, they do not alleviate the problem of double satellite hops, where absolute delay will be a source of customer dissatisfaction.

Conclusion

As noted at the beginning of this section, there are still incompatibilities between European and North American networks.

When the digital transmission hierarchy was first defined in the early 1960s, the industry adopted a format for the primary digital multiplex at 1.544 Mb/s. The Europeans later adopted another at 2.048 Mb/s. The differences propagated throughout the higher levels of the digital hierarchy. As a result, the world has had a North American and European market for digital systems ever since.

With the introduction of an integrated services digital network (ISDN), which will be covered in Section 18.1, the differences in North American and European standards are still creating problems. In North America, the achievable data transmission equivalent of a digital voice connection in many parts of the network is 56 kb/s, while in Europe it is 64 kb/s. Modern deployment strategies of digital transmission systems in North America take into account the need for 64 kb/s clear channel capability (CCC).

There is a commitment within various North American committees and the International Telecommunication Union – Telecommunication Standardization Sector (ITU-T) to reach agreements on these issues which will permit the development of a global ISDN.

Review Questions for Section 7

1. Briefly describe the direct distance dialing (DDD) network.

2. If −10 dBm is sent at the 0 dB TLP (transmission level point), what level would be read at the −2 dB TLP?

3. What is the maximum allowable round-trip delay in a circuit without echo suppression or echo cancellation?

4. In a digital network with analog subscriber loops, where is the major source of transmission impairments?

5. Describe the operation of a digital echo canceller.

6. At what distance does a circuit require some form of echo suppression using VNL?

8 The Local Subscriber Loop

Alexander Graham Bell issued a prospectus on the telephone in 1878 which stated:

Cables of telephone wires could be laid underground or suspended overhead, communicating by branch wires with private dwellings, country houses, shops, manufactories, etc., etc., uniting them through the main cable with a central office where the wires could be connected as desired, establishing direct communications between two places in the city.

This proposal foretells with remarkable accuracy both the structure and the scope of the customer loop that has been implemented over the last 100 years. The basic structure of the local loop is shown in Figure 8-1.

Figure 8-1   **The Local Loop**

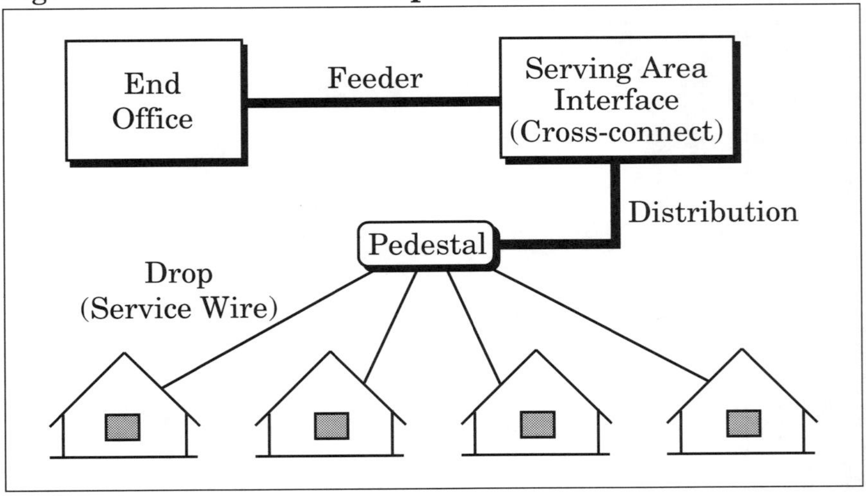

Today, the local loop represents an investment well in excess of $50 billion, the largest single asset of the local operating telephone companies. Annual expenditures on local loop facilities, for both growth and maintenance, total nearly $5 billion.

The telephone set is supplied with direct current (DC) from a 48-volt common battery supply in the end office over a pair of twisted copper wires. The two wires form a loop from the telephone set to the telephone office, so it is referred to as the local loop.

The batteries are backed up by a diesel generator at the local office. As a result, the telephone system can work even if there is an electrical outage.

The local loop provides the customer with access to the telecommunications network through the end office. In the United States, the end office connects with the interexchange carrier's point of presence (POP) either directly or through the telephone company's access tandem.

In the early days of telephony, uninsulated pairs of wire were strung on poles and called open wire. Today, the local loop consists of a pair of insulated wires twisted together and combined with hundreds of other twisted pairs into a single cable. These cables can be strung on poles, buried underground or installed in underground conduit. The diameter of the wire varies from 0.016 inch (26 gauge) to 0.036 inch (19 gauge): the thinner the wire, the higher the loss.

The transmission losses for the analog loop are expressed in decibels (dB) at a frequency of 1,000 Hz. The maximum loss allowed is 8.0 dB. The maximum loop resistance (1,300 ohms) is limited by a minimum loop current of 20 milliamperes (mA). The use of 26-gauge copper wire is restricted to 15,000 feet (4.6 km).

To reduce the loss or attenuation at voice frequencies in long local loops, inductors are placed in series at fixed intervals along the line (see Figure 8-2). Since these series inductors load the line to reduce attenuation, they are called loading coils. Their inductance is typically 88 millihenrys. A typical cable pair designation would be 22H88. This is 22-gauge copper wire with H indicating 6,000 feet spacing and 88 signifying the inductance. The design distance for 19-gauge to 26-gauge H88 loaded twisted pairs is shown in Figure 8-3. Another common loading scheme, called D66, uses 66 millihenry coils spaced 4,500 feet apart. The design distances for D66 are approximately the same as for H88.

Significant changes have occurred in the customer loop over the past two decades, as electronics began to play an increasing role. In the late 1960s and early 1970s, analog techniques were used to improve the performance of long customer loops and to eliminate the need for large (19-gauge) copper wire sizes on long loops.

In the late 1970s and early 1980s, digital subscriber carrier systems came into general use. The primary objectives of using digital carrier systems were to reduce the number of copper pairs required in the feeder portion of the customer loop and to limit the length of the distribution portion of the loop to 12,000 feet or less. Providing a remote terminal (see Figure 8-4) at the end of the feeder section, operating over a digital carrier system, reduces the amount of twisted pair cable required and eliminates the need for reinforcing feeder

Figure 8-2　**Loaded Line**

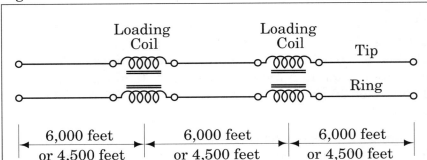

Figure 8-3   **Design Distance with 19 – 26-Gauge Wire**

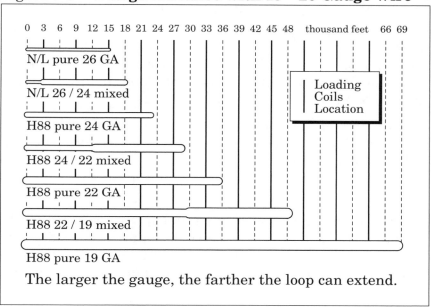

The larger the gauge, the farther the loop can extend.

Figure 8-4   **DLC Remote Terminal or**
Remote Switching Terminal

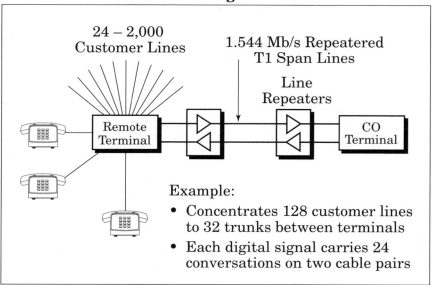

routes to accommodate customer growth. In addition, the use of digital carrier systems allow 64-kb/s or higher-rate digital services to be provided on even the longest loops. The steadily falling cost of digital carrier systems, added to the flexibility they offer in providing future digital services, has resulted in local telephone company standardization on the use of digital carriers on most loops of 18,000 feet or more.

Conditioning of the line (that is, the removal of bridged taps and loading coils to achieve the bandwidth necessary for digital transmission) and the installation of repeaters every mile are required. This cost is generally less than the cost of installing new cable.

In applications where use of existing copper pairs is not practical and installation of a new cable is required, it is becoming more economical to install fiber optic cable. As a result, nearly 95% of new digital carrier installations that require cable now use fiber.

Telephone companies are planning to provide fiber to the home (FTTH), or fiber to the curb (FTTC) in order to enter the video distribution market. This subject will be examined in Section 17.14.

However, in an effort to squeeze the maximum bandwidth out of the existing twisted pair loop, a number of new technologies have emerged. Asymmetric digital subscriber lines (ADSL), in combination with the latest advances in digital video compression, is now making it possible to transmit a digitally encoded video channel over the existing twisted pair loop. Video compression will be covered in Section 15. Telephone companies are exploring several ADSL options called ADSL-1, 2 and 3. ADSL-1 will operate over nonloaded loop plant which extends to 18,000 feet. It will operate at 1.5 Mb/s and deliver a single channel of encoded video. The quality level will be equivalent to a VHS tape played in a typical VCR. ADSL-2, operating at 3 Mb/s, will extend to 12,000 feet. ADSL-3, operating at 6 Mb/s, has been proposed for loops up to 8,000 feet (see Figure 8-5).

Figure 8-5 **Asymmetric Digital Subscriber Line**

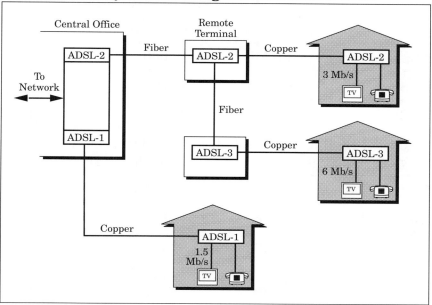

The existing local loops support switched and dedicated digital services at speeds up to 64 kb/s. The loop also supports the ISDN digital subscriber line (DSL), which provides two 64 kb/s "bearer" channels (circuit switched) and one 16-kb/s data channel (packet switched). The DSL can extend up to 18,000 feet from the central office or a digital loop carrier remote terminal on nonloaded cable. A number of commercially available pair gain devices use DSL technology to double or quadruple the capacity of a given twisted pair in the loop. These pair gain devices use 64-kb/s or 32-kb/s encoding, respectively, to achieve the pair gain.

The high bit rate digital subscriber line (HDSL) is being widely deployed in the local loop environment. The HDSL provides a standard DS-1 rate signal up to 12,000 feet from the serving central office using two cable pairs and no repeaters. The following table summarizes the distance and pair requirements for the various digital services that use twisted pairs.

Digital Loops

Name	User Payload	Max. Distance from CO	Pairs Required
ADSL-1	1.5 Mb/s	18,000 feet	1
ADSL-2	3 Mb/s	12,000 feet	1
ADSL-3	6 Mb/s	8,000 feet	1
Dedicated Digital Services (DDS)	2.4 to 64 kb/s	12,000 feet (more at the lower speeds)	2
Switched 56	56 kb/s	18,000 feet	1 or 2
ISDN DSL	144 kb/s	18,000 feet	1
Repeatered T1	1.536 Mb/s	up to approximately 200 miles	2
HDSL	1.536 Mb/s	12,000 feet	2

The existing twisted pair loop has served the public well for over 100 years, but broadcast quality video and other broadband services will require the provision of fiber directly to the home.

Conclusion

Although improvements in transmission technology have had a major effect on long distance transmission, most local loops still use twisted pairs. The local loop with its limited bandwidth is still adequate for voice communications and many digital services.

However, new services and enhanced equipment will dramatically change the requirements on the customer loop over the next decade. The provision of fiber to the residential customer will create a broadband integrated services digital network (BISDN). This concept will be examined in Section 19.

Fiber optics in the customer loop will change the way basic telephone and entertainment services are offered. The change will not be a revolution, however. For economical reasons, gradual evolution of the customer loop, making full, effective use of existing network facilities as well as available technologies, will be the rule.

Review Questions for Section 8

1. What is the maximum loss allowed in the analog local loop?

2. What is the maximum loop resistance in the analog local loop?

3. What is the minimum loop current in the analog local loop?

4. Explain the cable pair designation 22H88.

5. Through which office in the switching hierarchy does the local loop provide the customer with access to the network?

6. Do 19-gauge twisted pairs have a higher loss than 26-gauge twisted pairs?

7. How many customer lines can be concentrated on the remote terminal of a digital carrier system operating over two cable pairs?

8. At what bit rate does ADSL-1 operate?

9. What is the maximum operating distance for ADSL-2?

10. How many cable pairs does the ISDN DSL require and what is its maximum operating distance?

11. At what speeds do the dedicated digital services operate?

9 Microwave Radio

Microwave radio has been the backbone transmission system of the long distance telephone network in the United States and Canada for the past 35 years. Even with the introduction of satellites and fiber optics, microwave terrestrial radio continues to play a dominant role in communications networks throughout the world. Canada turned up its first coast-to-coast microwave radio system (TD-2) equipped with 180 voice circuits in 1958. The TD-2 and its variations were also heavily used in the United States for civilian and military long-haul applications.

Until the mid-1970s, nearly all such radio systems were analog, transmitting analog multiplexed signals originating from analog telephone switches. Although such networks provide good voice telephone service, even over long distances, they are less than satisfactory for data transmission at high speed and low error rates.

Digital transmission systems, however, detect and reshape the transmitted pulses at all repeaters as described in Section 3. This avoids the general buildup of noise which occurs on any long-haul analog system.

The world's first long-haul digital radio system, the DRS-8, was designed in Canada by Bell Northern Research (BNR) for Northern Telecom and has been in service since 1978. The DRS-8 has a capacity of 1,344 voice circuits and was developed for the 8 gigahertz (GHz) band. This system became the mainstay of the Canadian digital network.

The AT&T divestiture, together with the rapid development of new carrier networks by other common carriers, created a demand in the United States for digital radio systems which operate in the established common carrier bands at 4, 6 and 11 GHz. To be cost-effective, a digital radio system must transmit two DS-3 rate signals (corresponding to 1,344 voice channels) in each radio channel of a 4-GHz system and three DS-3 rate signals (corresponding to 2,016 voice channels) in each channel of a 6-GHz or 11-GHz system.

These high frequencies are conducted through metal pipes called waveguides before being transmitted over the air from the antenna. The radio beam follows a line-of-sight path, so a series of towers must be provided over the route of the system and spaced about 26 miles apart. An antenna on each tower receives the radio signals, amplifies (or regenerates, depending on the type of repeater) the signal and retransmits the signal to the next tower as shown in Figure 9-1.

Figure 9-1   **Radio Signal Path**

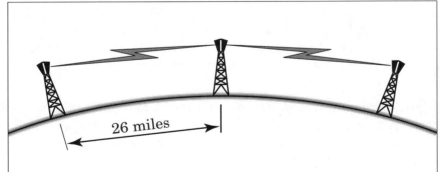

- Radio beams follow a line-of-sight path
- High frequencies are conducted through metal pipes called waveguides, which couple the radio to the antenna.
- At each tower the radio signal is received, amplified and transmitted to the next tower

60

Three major design considerations must be addressed by system planners. First, to avoid interference from other microwave radio systems operating in the same band, only one user can occupy a frequency band between the same locations in a given geographical area. This has created the need for users to coordinate the use of frequency or spectrum and the involvement of government regulatory agencies.

Second, as the microwave signal traverses the line-of-sight path in the atmosphere, it is subject to a variety of degradations. One of the most common degradations is multipath fading. Most of the time the microwave beam traverses the line-of-sight beam unimpeded. However, during hot summer nights when the air is still, distinct layers of warm and cold air may form along the path. The beams may then be directed to the receiving antenna along two different paths, a direct beam path and a refracted beam path. If these two signals add in phase opposition, the result will be a net loss in received signal. This type of loss is usually accompanied by a frequency-dependent phase and amplitude distortion in the received digital radio signal. If such distortion remains uncontrolled, signal outages can occur. Multipath fading, caused by interference from a beam reflected by the earth's surface, especially when the line-of-sight path crosses water or flat terrain, is another hazard to the microwave.

To overcome these problems, first-generation digital radio systems used a space diversity antenna, spaced 10 meters to 15 meters vertically below the main receiving antenna. The signals from the two antennas were then combined to reduce the amount of fading. A dynamic slope amplitude equalizer in the receiver then reduced distortion to an acceptable level. As a result, careful attention to design is critical if satisfactory performance is to be achieved.

An example of the effect multipath fading has on the received pulse signal is shown in Figure 9-2.

Figure 9-2   **Microwave Multipath Fading**

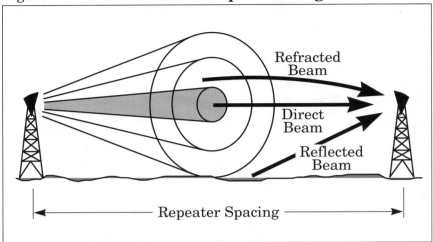

Third, in many cases, only existing paths, which were not originally designed for digital transmission, can be used, requiring the digital radio system to coexist with analog radio systems and antenna systems. Subject to regulatory control, the right of way for microwave is free, unlike an optical fiber cable system, and in many cases digital radio can be accommodated in existing towers and buildings originally intended for use with analog radio. A typical digital microwave radio system is shown in Figure 9-3.

Parabolic dish antennas are normally used for a microwave system operating in one frequency band, while horn antennas permit operation in several frequency bands, for example, in 4, 6 and 11-GHz bands using the same antennas.

In order to transmit a 90 megabits per second (Mb/s) signal (two DS-3s) in only 20 megahertz (MHz) of bandwidth, a modulation technique called 64 quadrature amplitude modulation (64 QAM) is used. If information is transmitted in a binary digital signal, each Mb/s of information would require about 1 MHz of bandwidth. To reduce the bandwidth required, a multilevel signal is used, where each signal element (or sym-

Figure 9-3   **Microwave Radio System**

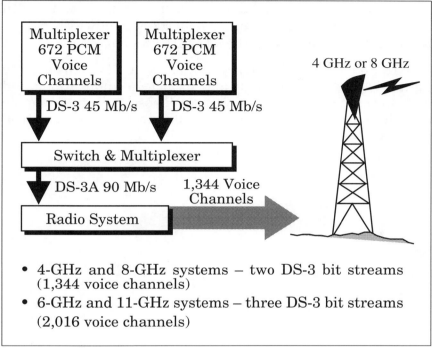

- 4-GHz and 8-GHz systems – two DS-3 bit streams (1,344 voice channels)
- 6-GHz and 11-GHz systems – three DS-3 bit streams (2,016 voice channels)

bol) received contains several information bits. In the case of 64 QAM, each signal element contains six bits. This and other multilevel encoding schemes provide higher bandwidth efficiency than simple binary schemes.

The latest development in microwave radio systems expands the bandwidth to match the land-based optical fiber systems which are based on the international synchronous optical network (SONET) standard. This standard is described in Section 17.8.

One commercial implementation of a SONET radio operates in the 4-GHz frequency band with 11 working channels and one protection channel, each 40 MHz wide. The system has a capacity of 44,000 voice circuits and will interface with the lightwave network at a rate of 622 Mb/s. This high-speed transmission rate was achieved by implementing a 512-QAM scheme. This higher level of modulation provides higher bandwidth efficiency, at 8 bits per second per hertz of bandwidth, rather than 6 bits per hertz with the 64-QAM scheme.

The 18 and 23-GHz frequency bands are used extensively by digital radios in short-haul and local loop applications and will continue to grow in the future. These radios use simple modulation systems, since occupied bandwidth is not a critical factor at these frequencies.

Conclusion

Digital microwave radio systems, together with lightwave systems and communications satellites, provide the major portion of the world's medium and long-haul transmission systems and will continue to do so in the foreseeable future.

While the North American market for microwave radio systems has peaked due to competition from lightwave systems, SONET radio will find many applications. The interconnectivity will offer telephone companies an alternative to lightwave systems for applications where radio is more cost-effective. In addition it will enhance the survivability of the lightwave network by providing a backup transmission medium in the event of a system failure.

SONET radio has been installed by several Canadian telephone companies to interconnect with lightwave networks. This 5,000 km network is the longest SONET-compatible network in the world.

Review Questions for Section 9

1. What is the major advantage of a digital microwave system over an analog microwave system?

2. How many voice circuits can each radio channel carry in an 8-GHz digital microwave system?

3. What is the average distance between digital microwave towers?

4. What are the three major design considerations that must be addressed by digital microwave system planners?

5. What is the difference between a parabolic dish and a horn antenna?

6. What advantages does a microwave system have over a fiber system when providing the right of way?

7. How many voice circuits can a fully equipped SONET radio system carry?

8. At what bit rate does SONET radio interface with the lightwave network?

10 Satellite Communications

The first sounds to be transmitted from outer space by a man-made device were "beeps" from the Russian *Sputnik I* launched on October 4, 1957. The United States entered the space age on January 31, 1958 with *Explorer I*.

The United States placed the first communications satellite in orbit a year later. *Score*, a short-lived but highly successful satellite, relayed messages up to 3,000 miles and broadcast to the world a tape-recorded Christmas greeting from President Eisenhower (see Figure 10-1).

The concept of a geostationary satellite was first published in October 1945 in an article entitled "Extraterrestrial Relays" by British scientist Arthur C. Clarke. Although his article seemed pure science fiction to many at that time, less than 20 years later the world's first commercial communication satel-

Figure 10-1 **Score**

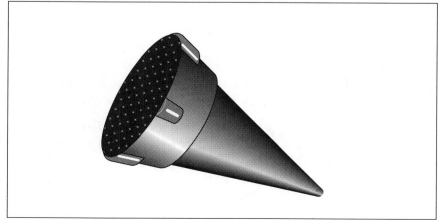

lite was launched on April 6, 1965, by the Communications Satellite Corporation (Comsat) in the United States. Comsat represents the United States in the International Telecommunications Satellite Consortium (Intelsat).

This synchronous satellite, named *Intelsat I* or *Early Bird*, was placed in geostationary equatorial orbit over the Atlantic Ocean. From its vantage point 22,300 miles above the Atlantic, *Early Bird* linked North America and Europe with 240 high-quality voice circuits and made live television commercially available across the Atlantic for the first time.

Within four years, six larger and more powerful satellites were successfully placed in operation over the equatorial regions of the Atlantic, Pacific and Indian Oceans.

Although the United States was a pioneer in the establishment of an international satellite network, Canada had the first domestic geostationary satellite system and continues to be a leader in this technology.

Telesat Canada, established as a Canadian corporation by an act of parliament on September 1, 1969, was authorized to install and operate a commercial satellite communications system throughout Canada. The system included a variety of communications repeater satellites in geostationary orbits, tracking, telemetry and command facilities and hundreds of earth stations of various sizes.

Telesat Canada launched its first satellite in November 1972, designated as *Anik A1*. In April 1973, *Anik A2* was launched, followed by *Anik A3* in May 1975. The following years saw a series of satellites, designated *Anik B*, *C*, *D* and *E* launched to carry telephone traffic and video services across Canada.

The launching of early communication satellites was an extremely costly operation using Thor-Delta and other series rockets. However, the reusable space shuttle, shown in Figure 10-2, is a manned vehicle that can place very large payloads in orbit more economically. The space shuttle is put into circular

Figure 10-2   **Space Shuttle**

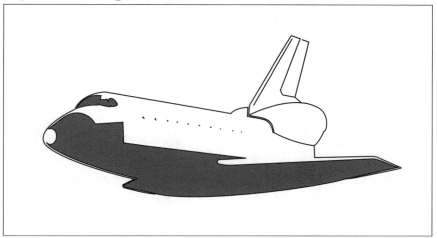

orbit at 320 kilometers (200 miles). To launch a satellite, a system is used to spin the satellite to 60 rpm and then eject it by means of springs. The satellite coasts clear for 45 minutes before the solid propellant motor is ignited.

To achieve orbital velocity, a satellite must be lifted above the atmosphere and propelled around the earth at a speed that will produce a centrifugal force equal but opposite to the gravitational force at that altitude. If the speed is too fast, the satellite will fly off into space. If the speed is too slow, the satellite will be pulled back to earth by gravity. Figure 10-3 illustrates the principle of orbital velocity. When velocity, direction and gravitational force balance, the satellite "falls" in a circular orbit. The earth's gravitational attraction decreases with altitude. Therefore, high-altitude satellites do not have to circle the earth as fast as low-altitude satellites.

At an altitude of 35,880 kilometers (22,300 miles), a satellite obtains a stable orbital speed of 6,870 miles per hour. At this speed, it orbits the earth in exactly 24 hours. This is a geosynchronous satellite because its orbit is synchronized to the earth's rotation. A synchronized satellite, placed in orbit directly above

Figure 10-3   **Orbital Velocity**

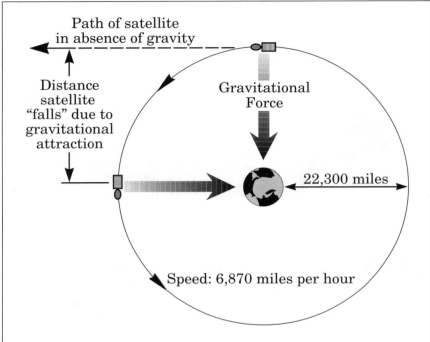

the equator on an eastward heading, appears to be stationary in the sky, so its orbit is referred to as a geostationary orbit.

Before the late 1970s, nearly all communications satellites used the 6/4 GHz band. The 4 GHz band was used for the downlink, and the 6 GHz band was used for the uplink. Used together they are called the C-band. Since the frequencies of the band coincide with those used for terrestrial microwave systems, their applications were limited, due to radio interference. However, the more powerful "second-generation" satellites operating in the 14/12 GHz band are capable of utilizing smaller and less expensive antennas. This band is referred to as the Ku band.

Satellite systems have to make use of the limited spectrum allocations. In addition, satellites must have sufficient spatial separation to avoid interfering with each other, so there is a limit to the number of satellites that can be parked in a given section of geosynchronous orbit. These satellites operate either in the C-band (6/4 GHz) or Ku band (14/12 GHz).

Under regulations established by the U.S. Federal Communications Commission (FCC) in 1974, U.S. communications satellites had to be positioned 4° apart (about 2,900 kilometers or 1,800 miles). As a result there were only 13 possible locations from which signals could be transmitted to the United States. The agency increased the total to 32 locations in 1977 by allowing the 14/12 GHz satellites to be positioned 3° apart. By requiring tighter control of transmitter power and antenna patterns, the FCC has since reduced the spacing to 2°, thereby increasing the number of allowable parking slots. Presently, there are almost 100 geosynchronous satellites worldwide. Covering North America, there are 14 C-band and 11 Ku-band satellites and an additional nine satellites with both C-band and Ku-band capacity.

The electronic circuitry on the satellite is called a transponder. It receives the signal transmitted from the earth station, amplifies the signal, changes the frequency and retransmits the signal back to earth. Each radio channel has its own transponder, so a number of transponders are on board the satellite to cover the allocated frequency band. Current communication satellites typically have 40 transponders, each with 36 MHz of usable bandwidth. A single transponder can carry one color television signal, 1,200 voice circuits or digital data at a rate of 50 Mb/s.

The size and shape of the satellite's radio beam on the surface of the earth is called the footprint. The actual antenna design and the transmitting power are important factors in determining the size of the footprint.

Signal transmission time or propagation delay and the associated echo become significant with satellite communications. The minimum distance between any two points via a satellite is 22,300 x 2 or 44,600 statute miles (72,000 kilometers). Consequently, a Los Angeles-Chicago circuit via satellite will have a round-trip delay of 540 milliseconds (ms) compared with the longest coast-to-coast terrestrial connection of about 50 ms (see Figure 10-4). Echo cancellers can control the echo, but nothing can be done about the delay.

With the widespread introduction of domestic satellite systems into the network, there is a possibility of satellite links being connected in tandem. Such double-hop satellite circuits should be avoided because the 1,080 ms delay (2 x 540 ms) is unacceptable to most people. While the delay problem is a major disadvantage for telephone circuits, satellites are ideal for one-way television transmission on transoceanic links and for domestic TV distribution.

Figure 10-4   **Transmission Time Delay**

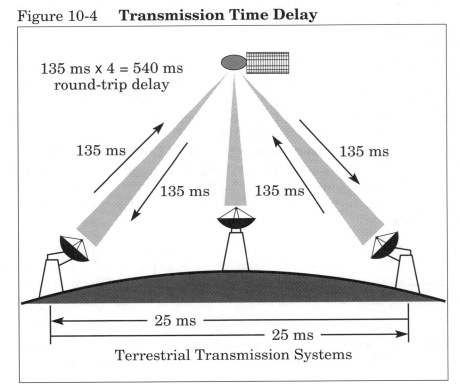

135 ms x 4 = 540 ms
round-trip delay

135 ms

135 ms

135 ms

135 ms

25 ms

25 ms

Terrestrial Transmission Systems

Ku-band earth station antennas typically are 1.8 meters to 2.4 meters in diameter. They are small enough to be located on the roofs of buildings without interference from existing microwave radio systems in the same area. The existence of the very small aperture earth terminal (VSAT) has made it possible to bring satellite communication directly to the end user's location.

There are a number of new nonswitched services integrating voice and data communications via satellite. They can operate between customer locations within one area or between that area and the rest of the country. These services use time division multiple access (TDMA) technology to carry the digital information. This technique allows more than one earth station access to a single satellite channel.

Using TDMA, the uplink signal from an earth station consists of a set of digital data bursts, each burst containing data addressed to a particular receiving earth station. The burst is assigned a time slot, and the earth station clocks are synchronized so that all are using a common set of time slots, one for each earth station in rotation. The entire rotation period may last less than 1 microsecond.

The receiving earth station receives all bursts, reads the addresses on the data, and processes only the data addressed to it. The voice services can be digitally encoded at 64 kb/s, 32 kb/s, or less. Data rates of 2.4, 4.8, 9.6, 19.2, 56 and 64 kb/s are available. The target market for this service is the medium to large-size business users who require a large portion of their voice and data traffic requirements to be dedicated to a private network.

NASA launched an advanced communications technology satellite (ACTS) in 1993. This multipurpose satellite offers experimental fixed, mobile and point-to-multipoint applications. It is capable of receiving signals at 30 GHz and transmitting them at 20 GHz.

Satellites for future systems will perform much more onboard signal processing compared to current satellites. They will have onboard switching and will be able to support inter-satellite links so that double satellite hops can be avoided. In addition, the satellite will act as a switch in the sky for major telephone links, directing traffic as it is received to the appropriate downlink.

Not directly related to telecommunications, but of importance to the timing and synchronization of both public and private telecommunication networks, is the global positioning system (GPS). GPS is used by military and civilian ships and aircraft for navigation with unprecedented accuracy to destinations anywhere on earth. This system consists of 24 Navstar satellites that circle the earth every 12 hours at an altitude of 20,200 kilometers. Four satellites are located in each of six planes as shown in Figure 10-5. This system has created a huge commercial market for hand-held GPS receivers. These units are used by weekend sailors and backpackers. A glance at the unit's display gives the traveler's latitude and longitude to within 30 meters, speed and course and the probable time of arrival at a preprogrammed destination. Drawing its power from flashlight batteries and using a small, built-in antenna, the unit takes about 30 seconds to find and lock onto radio transmission from at least three Navstar satellites (see Figure 10-6). The very high accuracy of the GPS timing signals is used in many telecommunication networks to supplement or replace the primary reference sources (PRS) traditionally used to synchronize digital networks.

In 1995, a new generation of satellites will be available. The Canadian MSAT satellite (see Figure 10-7) will be used in a joint U.S./Canadian venture to provide two-way communications between ground stations, vehicles, ships and aircraft. A twin MSAT, owned by American Mobile Satellite Corporation, will also be launched and the two MSATs will provide back-up to each other if necessary. A trucking company will be able to send information to a driver on the Trans-Canada Highway

Figure 10-5 **Global Positioning System**

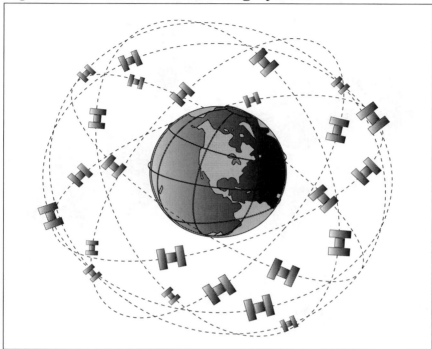

Figure 10-6 **Global Positioning System**

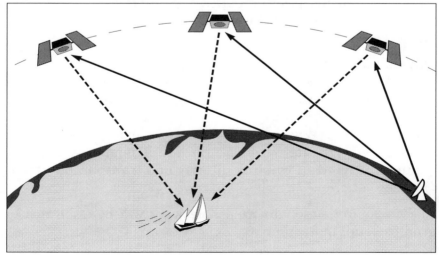

Figure 10-7   **MSAT Satellite**

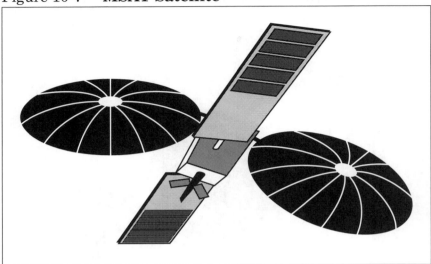

and another on the Los Angeles Freeway at the same time. This will create a nationwide system of mobile communications for American, Canadian and Mexican users.

Looking further into the future, the concept of medium Earth orbiting satellites (MEOS) and low Earth orbiting satellites (LEOS) is moving closer to reality. A joint venture of TRW and Teleglobe Canada will bring a system called Odyssey into global service in 1999.

Odyssey consists of 12 satellites operating at medium Earth orbit (MEO). This system will offer less voice delay and echo than geostationary systems. The Odyssey handset, essentially a palm-sized earth station, will operate in both cellular and satellite modes. Where terrestrial service exists, the Odyssey handset will operate normally. Where it is absent or interrupted, the handset will link directly and transparently to an Odyssey satellite.

The concept of a global wireless network will be examined in greater detail in Section 16.

Conclusion

Satellite communications are used extensively in telephony applications, particularly in sparsely populated areas of the world. The delay problem in satellite communications is a serious disadvantage for voice communications. However, satellites offer a wide bandwidth at a transmission cost that is independent of distance. As a result, they are very suitable for one-way television transmission between various network locations. Almost all network and cable television broadcasters receive most of their broadcast material from satellites.

A combination of satellite communications and cellular phones is moving closer to reality and will be examined in Section 16.

Review Questions for Section 10

1. What is a geosynchronous satellite?

2. At what altitude does the space shuttle launch a satellite?

3. At what altitude does a geosynchronous satellite obtain a stable orbit?

4. Briefly describe the function of a transponder.

5. What is the capacity of a typical transponder?

6. What is meant by a "footprint"?

7. What is the round-trip delay on a typical satellite link?

8. What is the advantage of a satellite operating in the Ku band compared to the C band?

9. What does VSAT stand for?

10. Briefly describe the function of time division multiple access (TDMA) technology.

11. How many satellites are used in the global positioning system?

12. At what altitude do the Navstar satellites operate?

11 Switching Systems

In 1889, a Kansas City, Missouri, undertaker, Almon B. Strowger, began to suspect that potential clients who called the operator of the local manual exchange and requested "an undertaker" were more often than not being connected to a firm down the street. This suspicion was reinforced when he learned that the telephone operator was the wife of the owner of the other funeral parlor in town.

As a result, Mr. Strowger invented a mechanical substitute for the biased operator that could complete a connection under direct control of the calling party. This simple device is called the Strowger, a two-motion, or step-by-step, switch. It was patented in 1891 and became the basis of a very large portion of the installed telephone switching systems in the world.

The Strowger switch system was manufactured by the Automatic Electric Company and sold to non-Bell independent telephone companies. The first installation of a Strowger switching system in the Bell system did not occur until 1917.

The Strowger, or step-by-step, switch connects pairs of telephone wires by progressive step-by-step operation of several series switches (called the switch train) operating in tandem. Each operation is under direct control of the dial pulses produced by the calling telephone.

The first switch in the train (the linefinder) searches for the line desiring service, the first and second selectors each take in one digit, and the last switch in the train (called a connector) takes in the final two digits.

A 10,000 line exchange requires four digits to be dialed (0000 through 9999). A typical Strowger switch bank is shown in Figure 11-1. Unfortunately the Strowger switching system generated large amounts of electrical and mechanical noise, it had high maintenance costs and a call could become blocked partway through the dialing sequence. The panel switching system was developed by the Bell System in 1921 as an alternative to the Strowger system. However, it was also extremely noisy and needed considerable maintenance. As a result, a new type of switching was introduced in 1938, called a crossbar switch. Crossbar, as the name implies, depends on the crossing or intersection of two points to make a connection. The basic crossbar switch is a matrix with 10 horizontal rows and 20 vertical columns as shown in Figure 11-2.

Therefore, any one of the input lines can be connected to any one of the output lines by energizing a particular input line and a particular output line. The early systems used electromechanical relays to provide the matrix interconnection. Later

Figure 11-1   **Strowger Step-by-Step Switch**

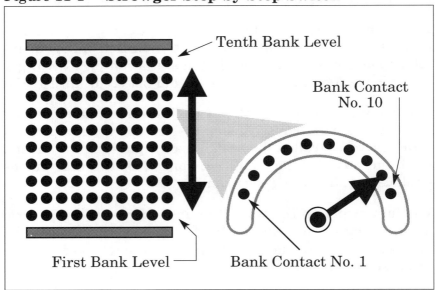

Figure 11-2   **Crossbar Switching**

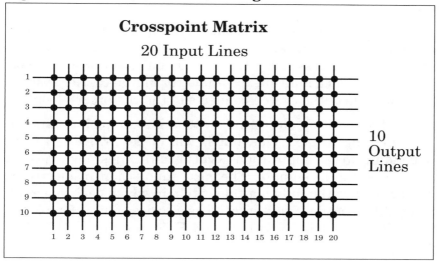

Crosspoint Matrix

20 Input Lines

10 Output Lines

models used reed relays because they were thought to be more dependable. The reed relay is a small, glass encapsulated, electromechanical switching device.

All step-by-step, crossbar and reed relay switching is called space division switching because each telephone conversation is assigned a separate physical path through the system.

The idea of using a programmable computer to control the operation of a switching network was a natural progression in the technology of switching systems. The first electronic system, the No. 1 ESS, was installed in 1965 and was designed to handle the switching for 10,000 to 65,000 lines. It had suffi-cient capacity to manage 100,000 calls per hour and a memory programmed with instructions on how to complete a call. The computer is kept aware of the status of the lines and trunks by means of the line scanners. They can detect a phone being picked up, the number being dialed and the completion of the call. They also detect the signals coming in from other switch-ing offices on the trunks. The computer controls the signals that are sent down the lines to the caller and to other switching offices, that is, dial tone, dialed number signals, rings and busy signals.

A typical electronic switch is shown in Figure 11-3. A trunk is a common communication channel and is terminated on the trunk side of the switch. This trunk could be combined with other trunks on a multiplexed analog or digital carrier system. The line side of the switch connects to the customer's loop.

All first generation electronic switching systems had an analog switching matrix. This matrix was either 2-wire or 4-wire, depending on the application (local or toll). These switches did not carry digital signals. The next innovation in electronic switching was the fully digital switch, in which each voice call is digitally encoded and switched as a series of bits. One of the first digital switches was the No. 4 ESS installed by AT&T in 1976.

A digital switch adds a new dimension to the space division technique, the dimension of time. The equivalent of a crosspoint in a space division matrix is a time-switched point or switchgate in the digital switch matrix. The switch is a solid-state device.

A digital switching system usually has both space division switching and time division switching. For example, a digital switch having a time division stage followed by a space division stage then another time division stage would be referred to as a TST switch, for time-space-time switch.

To switch these time division multiplexed circuits it is necessary to change the assignment of the time slots. The input to the digital switch might consist of a sequence A1/A2/A3/A4 which represents the digitized samples of four circuits. At the output of the switch, the new sequence might be A3/A1/A2/A4. If the output circuits are Y1 through Y4, then A1 is connected to Y2, A2 to Y3, A3 to Y1 and A4 to Y4. The switching is done by an interchange of the time slots using a time-slot interchange (TSI) unit, as shown in Figure 11-4.

TSI units are composed of many buffer memories. In a typical system each buffer memory can store 128 samples. The 128

Figure 11-3   **Electronic Switching System**

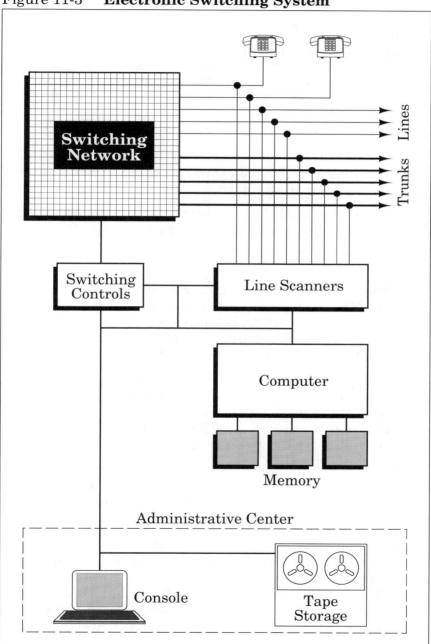

Figure 11-4 **Time Slot Interchange Unit**

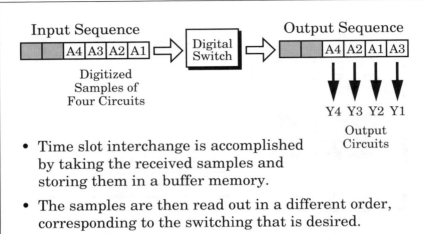

- Time slot interchange is accomplished by taking the received samples and storing them in a buffer memory.

- The samples are then read out in a different order, corresponding to the switching that is desired.

time slots are used for 120 digital circuits with eight slots for maintenance and spare capacity (see Figure 11-5). Eight buffer memories, with one spare, comprise a time slot interchange unit. A total of 128 TSI units are each connected at the input and output of a time multiplexed switch (see Figure 11-6).

Because each TSI unit has eight output lines, the time multiplexed switch must have 128 x 8 = 1,024 inputs and outputs. Each digital circuit is one-way, so two digital circuits are needed for a two-way circuit. The switch in this example can serve a total of 53,760 two-way circuits or 107,520 one-way trunks.

Digital switches are now available which allow the extension of switching or concentration modules to locations remote from the main switch location.

The main controlling portion of the digital switch is referred to as the **Base Unit (BU)** or host, and the remote portion located closer to the customer is referred to as the **Remote Switch Unit (RSU), Remote Switching Terminal (RST)** or the **Remote Line Unit (RLU)**. The links between the various units are PCM voice channels and a 64 kb/s digital data chan-

Figure 11-5 **Buffer Memory**

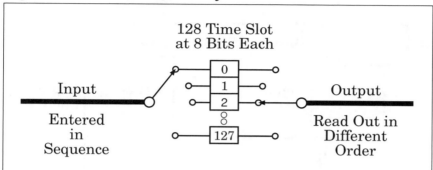

128 Time Slot
at 8 Bits Each

Input — Entered in Sequence

Output — Read Out in Different Order

- The 128 time slots are used for 120 digital circuits, with 8 slots for maintenance and spare capacity.

- 8 buffer memories, with one spare, comprise a Time Slot Interchange (TSI) unit.

Figure 11-6 **Digital Switching System**

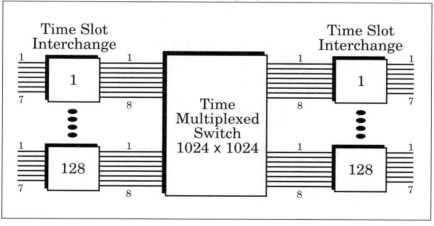

Time Slot Interchange

Time Multiplexed Switch 1024 x 1024

Time Slot Interchange

nel for communication between the processors in the base unit and the remote units. A typical distributed digital switch is shown in Figure 11-7.

Figure 11-7 **A Typical Distributed End Office**

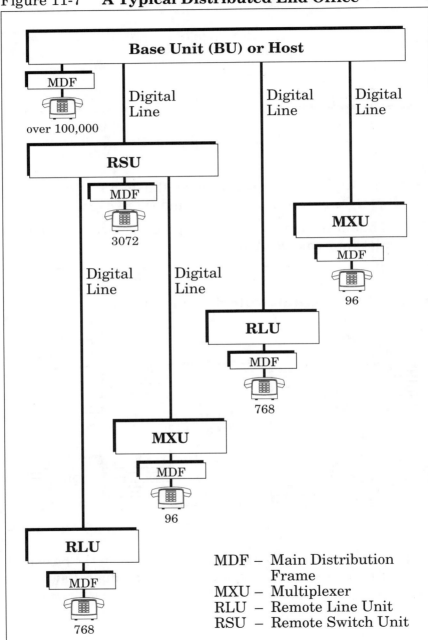

MDF – Main Distribution Frame
MXU – Multiplexer
RLU – Remote Line Unit
RSU – Remote Switch Unit

Conclusion

Along with digital techniques used in the customer's loop, the switch and the long distance trunk provide the highest quality service at the most cost-effective price at the present time. However, due to the rapid rate at which fiber optics is being installed in the toll network and the local distribution network, the electronic digital switch in its present form may have a short life. The slow speed of existing switching devices limits the benefits that can be obtained from sending information in the form of light through glass fibers.

When a lightwave system is switched from a trunk line to a local fiber distribution network, the message must be converted into an electrical signal, switched by a microelectronic device and then reconverted back to a light signal.

An optical or photonic switch would eliminate all those conversions. In addition, a future photonic switching system could support 10,000 channels, each operating at 150 Mb/s or more. The aggregate bit rate for this switching system would be greater than 1 million Mb/s or 1 terabit per second! In theory, this system would have the capacity to handle all North American voice traffic simultaneously.

Review Questions for Section 11

1. Briefly describe the Strowger, or step-by-step, switch.

2. List three disadvantages of the step-by-step switch.

3. How many input and output lines does a typical crosspoint matrix have?

4. What is meant by space division switching?

5. How many samples can each buffer memory in a typical digital switch store?

6. How many two-way digital circuits can a typical digital switching machine serve?

7. What is a remote line unit (RLU)?

8. In theory, what is the bit rate for an optical switch?

12 Private Branch Exchange (PBX)

Business organizations, in the early days of telephony, usually had a manual private branch exchange (PBX) installed on their premises. The PBX was, in effect, a telephone switching exchange acting as a branch of the local central office for the exclusive use of the organization.

When a call came into the PBX, the operator would determine the number of the called party, make the connection to the correct extension using a cord, and ring the number. Outgoing calls from the extensions would be connected through an appropriate trunk to the local central office.

Later, automatic PBXs were introduced utilizing the step-by-step and crossbar switching techniques. Most PBXs today are automatic, not manually switched by an operator. The term PBX applies to all types of private branch exchanges, manual or automatic.

A very simple form of PBX is the key telephone. In effect, it is a small station switching device. A typical key telephone has six keys, offering access to five lines and including the "Hold" key. The keys are lighted and flash in different patterns to indicate whether the line is in use, on hold, or ringing.

Today, like central office switching systems, there are two basic types of PBX: analog and digital. The basic difference between the two is the form in which the signal passes through the switching network. With an analog PBX the voice signal is passed through in its original analog form. Digital data to or from a terminal, computer or other digital device must be converted to an analog form to be switched through the analog

PBX. This is handled by a modem, which is described in Section 14. With a digital PBX, digital data are switched in digital form. Voice signals are first converted from analog to digital form by a codec in the telephone set or PBX station circuit and then switched.

One of the first digital PBXs was introduced in 1975. This represented a significant milestone in the evolution of business communications services. This system was the first digital switching system to use stored computer software control. The software provided access to additional user features that included call waiting, toll restriction, direct inward dialing (DID) and remote administration and maintenance. Digital PBXs have evolved over the years and now include special feature packages designed for large corporations, military forces, hotels, motels, hospitals and medical clinics.

A digital PBX is connected to the local central office by a number of circuits called direct-outward-dial (DOD) trunks. A typical digital PBX providing 256 digital voice and data stations would have 24 digital trunks operating over a T-1 carrier system. Small digital PBXs support as few as 30 stations, but are easily expanded in stages to meet new requirements as the business grows. The largest digital PBX has up to 60,000 stations.

A typical digital PBX has three main components; the central processing unit (CPU), the network and peripheral equipment, as shown in Figure 12-1. The CPU, following instructions stored in its memory, controls the switching function that connects PBX stations and trunks, which are part of the peripheral equipment. The operations, administration and maintenance functions can all be handled remotely from a system console.

The individual telephone sets can transmit data or voice, as well as access many advanced call processing features including autodialing, call forwarding, speed calling and ring again. In addition, more features are now becoming available. One feature is calling-name display (Figure 12-2), which provides the name associated with the calling party's telephone number.

Figure 12-1   **Digital PBX**

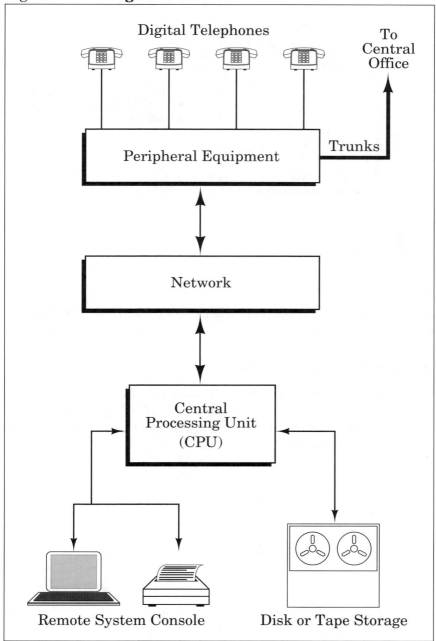

Figure 12-2   **Digital Telephone with Call Display**

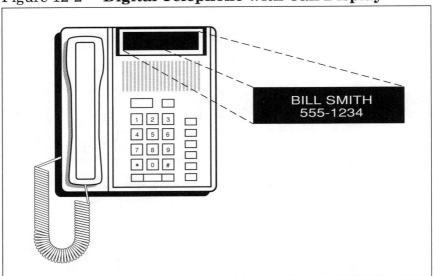

This feature can be coupled with the call-waiting feature and provides many businesses (stockbrokers, financial institutions etc.) with the ability to have client's records available on their computer screen as they answer the client's call.

As shown in Figure 12-3, the digital PBX can serve as a voice and data controller for the "automated office" by interconnecting computer terminals, word processing equipment, compressed videoconferencing, facsimile devices and telephones.

Figure 12-3   **Automated Office**

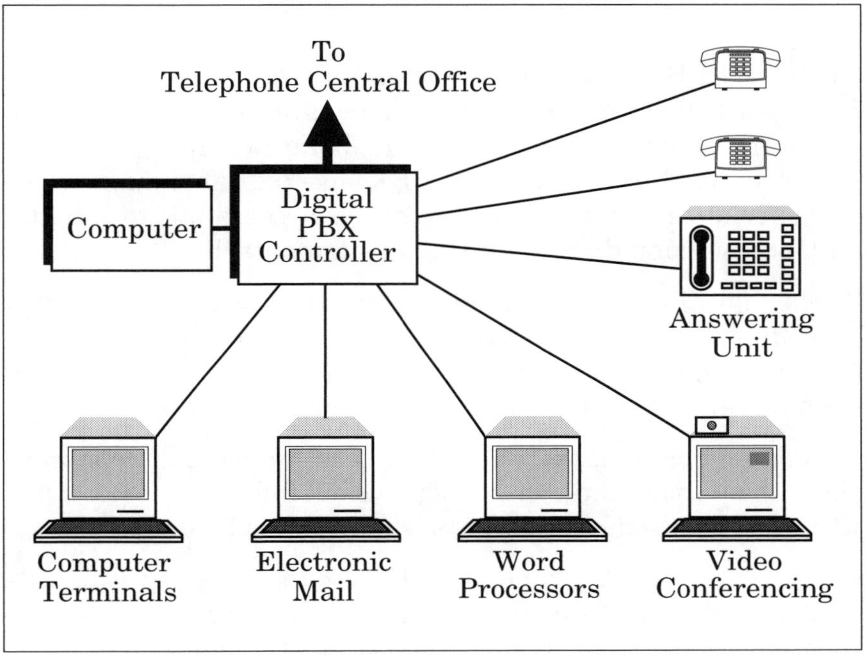

To
Telephone Central Office

Computer

Digital
PBX
Controller

Answering
Unit

Computer
Terminals

Electronic
Mail

Word
Processors

Video
Conferencing

Conclusion

The latest PBXs use similar technologies to those being used in network switching machines. They are essentially specialized minicomputers performing telephone switching for a business organization. All current PBXs use digital signals. They use special telephone sets that perform the analog-to-digital (A/D) conversion directly in the telephone itself.

These digital PBXs can interface with the local telephone central office over a digital T-1 carrier system or over analog trunks. In addition, they can serve as a controller for digital services such as time-shared computing and electronic mail.

In today's global marketplace, business success can depend on the cost-effective deployment of powerful information networks that link together a corporation's many worldwide locations.

Using 100% digital facilities, they can tie geographically dispersed digital PBXs into a single, unified corporate communications network providing voice, data and video services.

Review Questions for Section 12

1. What is the simplest form of PBX that can be used by a small business?

2. Explain the difference between an analog and a digital PBX.

3. Name the three main components of a typical digital PBX.

4. Explain the difference between a trunk and a station.

5. How many lines can the largest digital PBX support?

13 Traffic Considerations

The telephone systems in the United States and Canada handle over 600 million messages a day. These are routed over a comprehensive network of intercity trunks which interconnect more than 20,000 switching systems. This network serves all of the telephones in the two countries and provides connections to most other countries of the world.

Numerous algorithms have been developed over the years to predict how telephone traffic will occur, based on specific parameters. Telephone traffic is generally measured in terms of time, how many calls are made, and the amount of time required to make them. Traffic measurements include the total number of calls served during a specified time period. Usually, the specified time period is the busiest hour of the system, called the busy hour. Usage can be specified in terms of the percentage of time a trunk or switch is in use. Overflow can be measured as the percentage of calls that found some particular piece of equipment or the whole system busy.

Each local switching office is carefully designed to meet the needs of the customer area it serves. Sufficient call processing capacity is planned to allow roughly 10% to 12% of all customers to make a phone call at the same time. It is highly unusual for this capacity to be exceeded even during the busy hour, typically 10:00 AM to 11:00 AM (see Figure 13-1).

The CCS is a unit used to measure the volume of telephone calls. CCS stands for hundred (C = 100 in Roman numerals) call seconds per hour. The number of calls times their average duration in seconds gives the traffic in call seconds. Dividing

Figure 13-1   **Busy Hour Traffic**

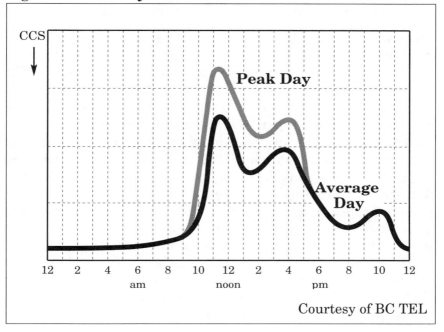

Courtesy of BC TEL

this figure by 100 gives the number of hundred call seconds (CCS). Another measure of traffic is the erlang, named after the Danish engineer and mathematician, A.K. Erlang. An erlang is equal to 36 CCS, or one hour. Depending on the requirements, you can use either CCS or erlangs in the calculations. To convert minutes to erlangs, multiply the number of minutes by 0.0167. To convert CCS to erlangs, divide by 36.

The unpredictability of telephone usage makes it necessary to build excess call-handling capacity into systems to ensure a high grade of service, typically measured in terms of the probability of call blockage. It also is necessary to include 15% to 25% excess capacity to account for false attempts and prematurely abandoned calls.

Mathematical models have been developed to describe the situation that results when a communication system becomes congested. These models relate the probability of blocking to the volume of traffic for different numbers of trunks.

The lower the probability, the higher the grade of service. Blockage is referred to in terms of a **P** grade of service. The **P** stands for Poisson, a French mathematician who developed the traffic algorithm that bears his name. The **P** is followed by a decimal number from 0.00 to 1.00. This figure refers to the fraction of calls that get blocked per 100 calls attempted. For example, if your telephone lines provide you with a P.05 grade of service, only about five out of every 100 calls you attempt will get blocked. This is considered a good grade of service within the industry, with P.01 being the best. Assuming a business has a private branch exchange (PBX) with total traffic during the busy hour of 50 calls at an average duration of five minutes per call, the total busy-hour traffic would be 50 x 5 x 60 divided by 100 = 150 CCS. The load factor can be determined by dividing the traffic (150 CCS) by 36 = 4. If the grade of service is P.01, then a PBX capable of handling four simultaneous calls would be sufficient for this volume.

Conclusion

Switching systems must be designed to handle the expected traffic. If a switching system is designed to be nonblocking, then all calls must be handled, regardless of the number of users served by the switch. Most switching systems block some of the calls, if they handle a large traffic volume.

Traffic engineering is an important part of the telecommunications business. Performing a traffic analysis and gathering all the pertinent data are the first steps in planning any telephone network.

Careful analysis of all traffic patterns and accurate forecasts of circuit growth are necessary in order to design the best network for the lowest cost.

Review Questions for Section 13

1. When does the peak calling time occur in the telephone network?

2. Call processing capacity is usually provided to allow what percentage of customers to make a phone call at the same time?

3. Describe the CCS unit of measurement.

4. How many CCS is equal to an erlang?

5. Describe a P.05 grade of service.

6. What is the total busy-hour traffic measured in CCS if a PBX is handling 50 calls at an average duration of five minutes per call?

7. If a P.01 grade of service is specified, how many simultaneous calls could be handled by the PBX?

14 Data Transmission

Data transmission actually began around 1835 with the invention of the electric telegraph and the development of the Morse code by Samuel Morse. The telegraph sent dots and dashes along a wire using electromechanical induction. The dots and dashes were used in various combinations to represent binary codes for letters, numbers and punctuation. The first telegraph printer was developed in 1849, but it was not until 1860 that high-speed printers operating at 15 b/s (bits per second) were available.

The first computer was developed by the Bell Labs in 1940 using electromechanical relays. The first mass-produced electronic computer was the UNIVAC built in 1951 by Remington Rand Corporation. In 1964, International Business Machines Corporation (IBM) introduced the System 360 line of mainframe computers. The first personal computers were introduced in 1977 by Apple, Radio Shack and Commodore. More than 5 million of these machines were ultimately sold.

With the proliferation of mainframe computers, small business computers and personal computers, the growth of data communication networks has ben dramatic. A typical example of a multipoint communication network would be a bank with many branch locations in a city, all tied into one main computer.

Digital computers understand only binary numbers, which are represented by combinations of the two digits **1** and **0**. Each **1** and **0** is called a bit (binary digit). Codes have been developed to translate the letters (A–Z), the numbers (0–9), and punctuation marks into binary **1**s and **0**s that computers understand. Each character is represented by a unique combination of a fixed number of bits, usually six, seven or eight, depending upon the particular code which is used.

A modem, which is an acronym for modulator-demodulator, is the interfacing device that couples the output of the digital data to the telephone lines. The sending modem converts the digital data to an analog format that can be readily handled by voice-grade telephone circuits, and the receiving modem changes the format back to the original signal (see Figure 14-1).

Computers use parallel transmission internally, which means all code elements are transmitted simultaneously. If an 8-bit word is transmitted, eight pairs of lines would be required. As a result, parallel transmission is only used within a computer center, for example, between a computer and a printer. The parallel transmission must be converted to a serial bit stream before it is coupled to a modem for transmission over the telephone network. This conversion device is called the universal synchronous / asynchronous receiver / transmitter (USART), as shown in Figure 14-2.

Figure 14-1   **Modem**

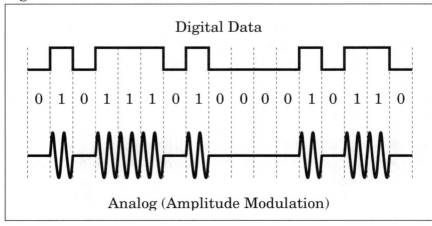

Digital Data

0 1 0 1 1 1 0 1 0 0 0 0 1 0 1 1 0

Analog (Amplitude Modulation)

Figure 14-2   **Parallel-to-Serial Conversion**

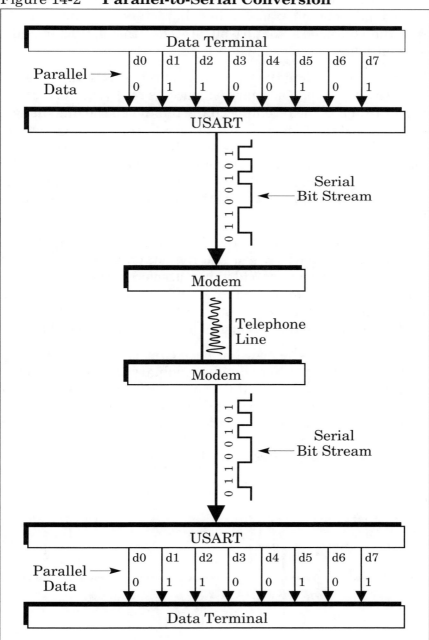

In asynchronous serial transmission each character is transmitted independently of all other characters. In order to separate the characters, a start bit precedes the character and a stop bit follows it, as shown in Figure 14-3.

In the idle condition, the transmission remains in the binary **1** state (mark). When a binary **0** (space) is received, the receiver begins to detect the incoming character. When the line changes from a constant binary **1** to a binary **0**, that is the start bit. When the receiver senses the start bit, it starts a clock which measures bit times. It then samples the next eight bits (7-bit character plus parity). The next bit is a stop bit, which must be a binary **1**. The parity bit (the eighth bit) is used to check for errors.

Parity checking counts the number of binary **1**s in the character and sets the parity bit so that the total number of binary **1**s is either odd or even. If odd parity is selected, the parity bit is set to binary **1** to make the total number of binary **1**s odd. If the character has an odd number of binary **1**s, the parity bit is set to binary **0**, so the number of binary **1**s remains odd.

If even parity is selected, the parity bit is set to binary **1** or binary **0** to make the total number of binary **1**s an even number. Referring to Figure 14-4, if the letter **"K"** was changed to the letter **"C"** by an error in transmission, the parity bit would still be set to binary **1**so the receiver would detect an error and

Figure 14-3   **Asynchronous Transmission**

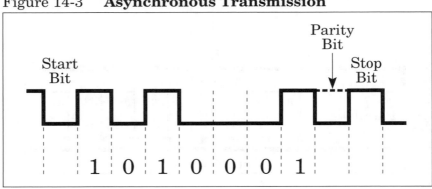

104

send a special parity error character. Transmission of the parity bit only allows somewhat limited error detection and does not allow error correction.

Synchronous transmission is used for high-speed transmission of a block of characters (see Figure 14-5). Synchronization is established by preceding the data block with synchronization characters. The sending modem clock controls the timing of each bit in the block of characters sent, including the synchronization characters. The receiving modem clock uses the

Figure 14-4   **Error in Transmission**

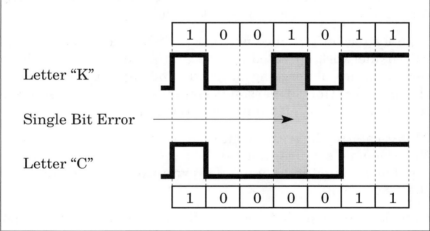

Figure 14-5   **Synchronous Transmission**

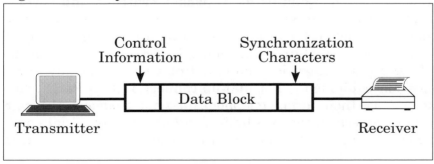

synchronization characters to establish clock timing for the rest of the block. Each block is also terminated by control characters.

In the past, data had to be given an analog "voice" (using tones) transmitted on the telephone lines. This is still done today in many cases, but it is a slow and awkward way to move bits. Human speech is therefore being digitized instead so that people and computers can use digital transmission facilities. Unlike voice conversations, which are continuous, transmissions of digital data tend to be "bursty." In response, "packet-switched" networks are being added to new end-to-end "circuit-switched" digital networks. Packets tie up the network only during the short time that they transmit.

A public packet-switched network (PPSN) is a data communications network operated by a local exchange carrier. These networks are undergoing dramatic and rapid growth due to the proliferation of personal computers and point-of-sale terminals, such as electronic cash registers.

This escalating interest in data communications, coupled with the economic savings that can be achieved by packet switching, has resulted in an annual growth rate in some networks that exceeds 40%.

Canada's public packet network, Datapac, was one of the first to be built. Established in the early 1970s, it is among the largest and fastest-growing public packet networks in the world.

Each PPSN has one or more packet switches and several access concentrators. Figure 14-6 shows a typical configuration. The network can be accessed by a customer via direct access lines or dial-up connections. Direct access lines may be of two types, using voice / data multiplexing or data only transmission equipment. For dial-up connection to a PPSN, customers use their normal telephone lines and modems. Although dial-up access presently is limited to 28.8 kb/s, voice / data multiplexed access is capable of data speeds up to 56 kb/s or, in

Figure 14-6   **Public Packet Switched Network (PPSN)**

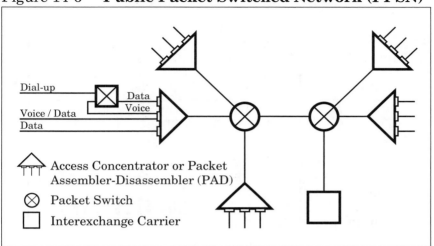

Dial-up
Data
Voice / Data   Voice
Data

Access Concentrator or Packet
Assembler-Disassembler (PAD)

Packet Switch

Interexchange Carrier

many networks, 64 kb/s. The packet switch can also be con-nected to interexchange carriers via separate digital channels.

The adoption of ITU's X.25 standard in packet networks made it easier to interconnect the services and equipment of differ-ent vendors. X.25 provides a set of guidelines that describe the physical interface between the terminal equipment and the communication network.   Many companies now use a hybrid X.25 network combining a private network with a public X.25 network. The X.75 protocol is used for communications with interexchange packet carriers.

X.25 packet switching is a technology of the mid-1960s for solv-ing the problems of scarce bandwidth and error-prone trans-mission media, with error control performed at every link. Frame relay is an evolutionary step beyond X.25 for improving packet network throughput by eliminating error checking or correcting on a link-by-link basis.

Frame relay is a system that transfers, or relays, frames of user data. Error checking is done only at each end. As shown in Figure 14-7, in the time division multiplex (TDM) frame,

Figure 14-7   **Packet Switching vs TDM**

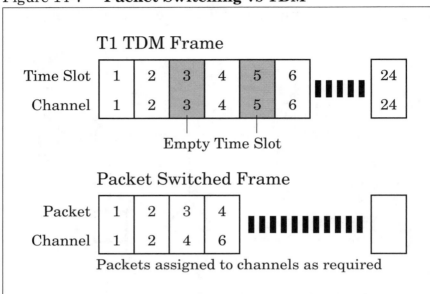

T1 TDM Frame

| Time Slot | 1 | 2 | 3 | 4 | 5 | 6 | ▮▮▮▮▮ | 24 |
| Channel | 1 | 2 | 3 | 4 | 5 | 6 | | 24 |

Empty Time Slot

Packet Switched Frame

| Packet | 1 | 2 | 3 | 4 | ▮▮▮▮▮▮▮▮▮▮ | |
| Channel | 1 | 2 | 4 | 6 | | |

Packets assigned to channels as required

each channel for which bandwidth is allocated occupies a pre-determined position in the bit layout of the frame, regardless of the instantaneous bandwidth requirement. The maximum bandwidth is permanently preallocated to the channel even when a channel is inactive and the time slot has no data.

In packet switched systems, bandwidth is allocated dynamically on demand, and the channel will use only as much bandwidth as needed and only when required for information transfer. When there are no data being sent or when a voice circuit is silent, no bandwidth is used. Giving users access to the whole channel at random intervals for random lengths of time, frame relay does not identify a given user's data on the basis of its particular time slot within a frame as TDM does. Therefore, frame relay requires a header containing an address representing the destination of the message.

Frame relay can be set up as a permanent virtual connection (PVC) or a switched virtual connection (SVC). A PVC is a nailed-down, high-speed channel that is established at provisioning time through the frame relay network. A PVC is maintained until administratively taken down. A SVC is a channel that is established through call setup procedures on a demand basis. The network establishes a virtual circuit to the specified destination and maintains the connection until the user terminates it.

As mentioned earlier, frame relay, unlike X.25, does not perform error checking on each link. This is left to the user equipment at the beginning and the end of each connection. By reducing the protocol processing overhead, frame relay may be less efficient than X.25, on error-prone links. However, with high-quality optical fiber circuits becoming commonplace in the public networks, error rates have been reduced by several orders of magnitude, and frame relay provides an economically attractive method of transmitting large volumes of data.

Frame relay is also more efficient because it does not require the call setup phase required for X.25. Delay through an X.25 public data network can be 200 ms or more. Frame relay can reduce the delay to about 20 ms. When frame relay is provided over permanent virtual circuits, no call setup is needed on a per packet or session basis since the address fields are agreed upon when the user subscribes to the service.

As shown in Figure 14-8, the frame relay packet consists of the variable length information field which can contain from 64 to over 1500 bytes or octets. The packet includes a two-octet address field, a two-octet frame check sequence and opening and closing flags of one octet each. The address field may optionally be extended to three or four octets. The information field consists of an integral number of octets (no partial octets). The flags indicate the beginning and end of a frame. The frames are transported transparently through the network. Only the address and frame check sequence fields may be modified. The data link connection identifier (DLCI) in the address field specifies a connection between two endpoints in the network and identifies all data on one preestablished path (virtual circuit).

Figure 14-8 **Frame Relay Packet Structure**

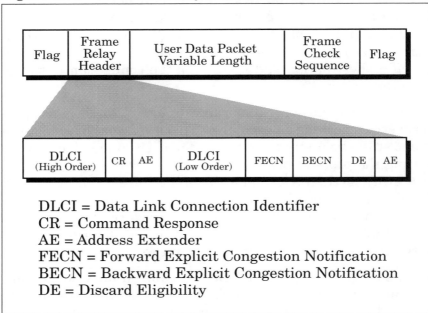

DLCI = Data Link Connection Identifier
CR = Command Response
AE = Address Extender
FECN = Forward Explicit Congestion Notification
BECN = Backward Explicit Congestion Notification
DE = Discard Eligibility

The same DLCI number could be used to identify an entirely different path between any other two endpoints.

The frame check sequence is a simple 16-bit cyclic redundancy code (CRC) check of the frame contents. If the frame is corrupted, a frame relay switch discards the packet, leaving the responsibility of requesting retransmission to the equipment at the endpoint as is done with TDM transmission.

The forward and backward explicit congestion notification (FECN and BECN) bits communicate a congestion indication. The discard eligibility (DE) bit indicates whether the frame should be discarded for rate control in the event of congestion.

Frame relay can operate at speeds of 56 kb/s to 1.5 Mb/s with an increase to 45 Mb/s planned for the future. Frame relay can support applications requiring medium to high-speed data transmission, image transfer and other bursty traffic which is relatively delay insensitive. It is less suited to delay-sensitive applications such as voice and video. Frame relay is ideal for applications that benefit from fast response time such as financial trading transactions, airline reservation systems and point-of-sale inquires. The prime application for frame relay service is local area network internetworking.

Several commercial implementations of frame relay are available in the United States. One particularly important frame relay service, known as switched multimegabit data service (SMDS) provides wideband services to businesses in many metropolitan areas. SMDS is a wide-area, public packet switched data service at DS1 (1.5 Mb/s) and DS3 (45 Mb/s) rates over fiber optic lines. It is considered a predecessor to the broadband integrated services digital network (BISDN).

Conclusion

Any home or office computer can use the telephone network to access another computer or reach most databases through the use of a modem. The speed of the communication can be from 300 b/s to 28.8 kb/s. The circuit-switched telephone network is suitable for slower-speed data communications and is quite economical for infrequent data traffic.

A frame relay packet switched network is more appropriate for business customers who want instant access to a dedicated network. The packets can be switched and transmitted over a number of different paths to their final destination at data speeds up to 1.5 Mb/s.

An all-digital network will present exciting possibilities. Over the next few years the BISDN will move closer to reality. This concept will be discussed in Section 19.

Review Questions for Section 14

1. What is the difference between a codec and a modem?

2. Is parallel transmission always used to connect a computer and a printer?

3. Describe the main function of a USART.

4. What is the purpose of the parity bit?

5. What is the difference between asynchronous and synchronous operation?

6. Briefly describe a packet switched digital network.

7. What is the name of Canada's public packet switched network?

8. Dial-up access to a PPSN is limited to what bit rate?

9. Voice/data multiplexed access to a PPSN is capable of transmitting up to what speed?

10. What ITU standard defines the interface between the packet-mode user device and the PPSN?

11. How is error checking done on X.25 and frame relay?

12. What is a PVC and a SVC?

13. What is the maximum size of the information field in a frame relay packet?

14. What is the highest operating speed for frame relay at the present time?

15 Video Transmission

When AT&T introduced its Picturephone at the New York World's Fair in 1964, the new technology generated great public interest. But the 1964 Picturephone was only an engineering demonstration. Actual commercialization of the technology was not then attainable for technical and economic reasons. However, many new products and services have entered the marketplace since 1964.

The major problem has been the need for an extremely broad bandwidth to carry the tremendous load of information in a video signal. When a standard monochrome television picture is digitized, each image is sampled over a grid of 512 horizontal by 480 vertical picture elements called pixels. Each pixel represents the brightness information and is usually coded with eight bits for a total of 256 possible shades of grey. To display motion, the picture is sampled at 30 frames per second. As a result, the rate is about 59 Mb/s (512 x 480 x 8 x 30). Adding color information brings the data rate to 90 Mb/s.

One solution to this situation was the creation of advanced video bandwidth compression techniques. Bandwidth compression, or more correctly, "bit rate reduction," significantly reduces the data rate needed to reconstruct the video signal at the receiving end. Using an elementary form of compression, such as sampling the picture at 15 frames per second, still requires about 45 Mb/s.

Video compression devices, known as codecs, provided the key to video transmission and represent a significant technological achievement. The video codec digitizes and compresses the analog video signals.

However, it became apparent that widespread acceptance of visual telecommunications as a work and management tool would not come until video could be transmitted over a nationwide network providing the convenience and economy of ordinary dial-up telephone service.

As a result of the move toward the integrated services digital networks (ISDN), most major telephone companies in North America are now offering a wide variety of high-speed switched data services, all with flexible access and competitive price structures. This has created a digital wide area network (WAN) with the same reliability and flexibility as the public-switched voice network. AT&T Communications' Accunet® Switched 56 Service is providing 56 kb/s service on a dial-up basis for intercity applications in nearly 400 cities. The service has been expanded to include Switched 64 (64 kb/s), Switched 384 (384 kb/s) and Switched 1536 (1,536 kb/s) service. Many of the Bell operating companies are also providing their own switched 56 kb/s service from their end offices. Both MCI and Sprint offer Switched 56 service as well as the higher-speed switched digital services.

However, to progress from 45 Mb/s video transmission to an acceptable 56 kb/s video transmission, a major breakthrough in image compression was required in order to provide a compression ratio of 800 to 1.

Powerful video codecs are available which can operate over 56 or 64 kb/s networks using new compression techniques. These techniques generally involve a combination of several basic schemes for picture compression. They utilize the correlation in space and time for video signals. Compression in space is known as intraframe coding, while compression in time is called interframe coding. A typical codec is shown in Figure 15-1.

One scheme squeezes out any information that is not absolutely necessary to recreate the picture at the distant end. The codec will predict what the image will look like from one instant to the next. Then it compares the actual image with the predic-

Figure 15-1   **Typical Codec**

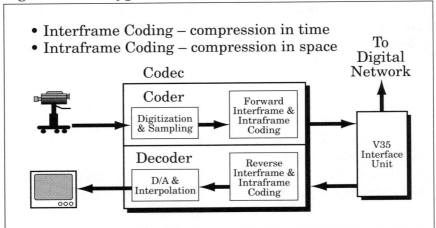

tion and transmits only the difference between the two. Since there is not much movement during a "head-and-shoulders" video teleconference, there is not much "difference" to transmit. The codecs are available in either the North American 525 lines NTSC standard or the European 625 lines PAL standard.

The higher the data rate, the sharper the picture quality. The most popular rate is 112 kb/s using two Switched 56 lines. This is not broadcast quality video, but it is perfectly acceptable for videoconferencing where the images are usually relatively static on the screen. Some "smearing" and freeze-frame effects are apparent with these relatively low transmission speeds.

Developments in the video teleconferencing arena have brought the prospect of videotelephones closer to reality. A number of major telecommunication companies are currently marketing both stand-alone videophones as well as some which will be embedded in personal computers.

A videophone, as a stand-alone terminal, is usually equipped with a 3 inch by 3 inch color liquid crystal screen, a video camera and a codec. In addition, it incorporates the features of a handsfree digital telephone set. When connected to ISDN, it will support the simultaneous transmission of image, speech and data. A typical videophone is shown in Figure 15-2.

Commercially available products range from a software only system which provides for interactive annotation, file and graphical image transfer to full motion video conferencing. Also available are advanced PC-based communications systems. More advanced systems include a video-capture board which can be plugged into a video camera or VCR and used to transmit still-frame images. Finally, a full motion package is available, which includes the software, board and a codec. It provides full color video conferencing at 30 frames per second over switched 56, 64 or 384 services offered by most telephone companies. A typical system is shown in Figure 15-3.

This technology is known as "Multimedia" and integrates the worlds of video transmission and digital communications with the world of computing. Companies such as Apple Computer,

Figure 15-2   **Typical Videophone**

Figure 15-3   **Multimedia System**

IBM, Intel, Microsoft and Texas Instruments have invested millions of dollars in multimedia. To succeed, multimedia requires a mass market and a high degree of standardization in the computer industry. A number of industry standards are emerging, one for photo-quality still images developed by the Joint Photographic Experts Group (JPEG) and another for full-motion images developed by the Moving Pictures Experts Group (MPEG). Until the development of MPEG, JPEG was used for full motion video. Running at 30 frames a second, it does not keep redundant data between frames and allows compression ratios as high as 50:1 without apparent loss of image quality. MPEG uses interframe compression and achieves compression ratios of about 50:1 by storing and transmitting only the differences between frames. While a digitized video image from a VCR or camcorder requires a data rate of about 27 Mb/s to transfer a screen full of pixels with 24-bit color running at 30 frames a second, MPEG reduces this to about 550 kb/s. The move to multimedia computing can be handled by existing 32-bit 386 DX and 486 DX CPUs. Using a new bus standard called VL-BUS, it can transfer data at up to 67 Mb/s. Another new bus standard, PCI, can handle 128 Mb/s and is based on Intel's existing 32-bit 486 DX and its 64-bit successor, the Pentium.

Multimedia computers are becoming more affordable. Most sell for less than $2000 with add-on upgrades such as a Compact Disc Read Only Memory (CD ROM) for an additional $200 to $600. There is already a large variety of material available on CD ROM including encyclopedias, atlases and books. Many claim that interactive multimedia will speed learning because one-on-one instruction means a student can master the material before moving on. Multimedia computers providing text, graphics, still images, sound, full motion video and video conferencing will offer professionals a powerful business tool. However, the success of the multimedia market will depend on how consumers prefer to receive information and entertainment, and the integration of computers and television sets.

Again, the rapidly evolving digital technology makes all this possible. The convergence of video, telecommunications, and the computer has finally made multimedia available to everyone.

Conclusion

Thirty years after AT&T introduced its Picturephone, the video-phone market is poised for rapid expansion. The long-awaited benefits will finally bring productivity gains to both large and small businesses.

As a result of product innovations, low-priced switched digital transmission services and a plan for standards, many organizations are seriously exploring the widespread use of videoconferencing. For the majority of videophone applications, dial-up switched service, typically at 112 kb/s (using two Switched 56 lines), is both convenient and economical. By using Switched 56 services and integrated videoconferencing systems, users can place a video call as easily as a conventional phone call.

As the market moves ahead, videophones and multimedia computers will become a standard addition to every company's communications, much as telephones and fax machines are today.

Review Questions for Section 15

1. What bit rate is required to digitize a black-and-white TV signal?

2. What bit rate is required to digitize a full-color TV signal?

3. What is the name of the device that compresses the bandwidth of a video signal?

4. Explain the difference between the North American NTSC standard and the European PAL standard.

5. Which bit rates are offered on commercial frame relay services?

6. What compression ratio is required to reduce a 45 Mb/s video signal to 56 kb/s?

7. Define "multimedia."

8. Name the industry standard for photo-quality still images and full-motion images.

16 Cellular Mobile Telephone Service

Cellular mobile telephone service is a system for providing direct-dial telephone service to mobile vehicles or personal communications devices, by using radio transmission.

Mobile telephone service has always been very popular with business customers, but its growth has been restricted due to the limited number of radio channels available. An entirely new approach utilizing a cellular concept was developed to provide high-quality mobile service for more customers at an affordable cost. Cellular radio has proven to be one of the fastest-growing technologies in the world.

The basic concept of a cellular system is frequency reuse, in which the area covered by the transmitter is reduced by reducing the transmitter power. In this way, concentrated areas of population can have more transmitting stations and thus more channels, because each transmitter handles a given number of conversations. In addition, because the lower-power transmitters cover less area, the same frequency can be reused in a common geographical area.

The basic system arrangement is shown in Figure 16-1. The service area is divided into regions called cells, each of which has equipment to switch, transmit and receive calls to / from any mobile unit located in the cell. A typical cell has a radius of 1 to 12 miles. Each cell transmitter and receiver operates on a given channel. Each channel is used for many simultaneous conversations or calls which are not adjacent to one another, but are far enough apart to avoid excessive interference. Thus, a system with a relatively small number of

Figure 16-1   **Basic System Arrangement**

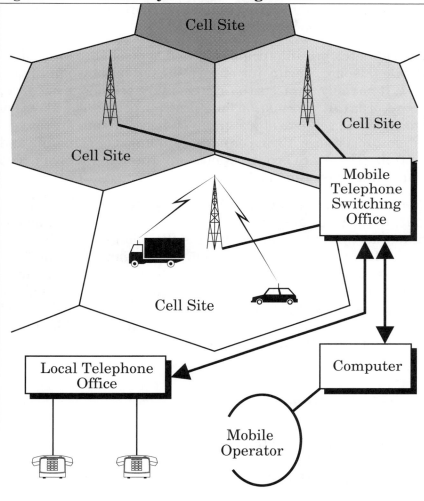

customers can use large cells, and as demand grows the cells can be divided into smaller ones.

This process of handling additional growth is called cell splitting. This is based on the fact that cell sizes are not fixed and may vary in the same area over time. Initially, all the cells in an area may be relatively large, but as the number of users in some cells increases, the overloaded cells are split into smaller cells by adding more transmitters.

The cell sites are interconnected and controlled by a central mobile telephone switching office (MTSO), as shown in Figure 16-1. It not only connects the system to the telephone network, but records call information for billing purposes. The MTSO is linked to the cell sites by a group of voice circuits for conversations, together with one or more data links for signaling and control. The MTSO controls not only the cell sites, but also many functions of the mobile units.

The mobile unit, shown in Figure 16-2, consists of a control unit, transceiver and antenna. The transceiver contains circuits that can tune to any of the 832 FM channels in the 800–900 MHz range assigned to the cellular system. The band from 824–849 MHz is used to receive signals from the mobile units and the band from 869-894 MHz is used to transmit signals to the mobile units. Each cell site has at least one setup channel dedicated for signaling between the cell and its mobile units. The remaining channels are used for conversations.

Figure 16-2   **Mobile Unit**

- Transceiver tunes to 832 FM channels – 800 to 900 MHz range
- Transmit channels – 824 to 849 MHz
- Receive channels – 869 to 894 MHz.
- Mobile unit transmits at 3 watts or less.

The base station typically transmits with a power of 25 watts, and the mobile unit transmits with a power of about 3 watts maximum. All first-generation cellular systems use analog transmission techniques. Frequency modulation (FM) is used for radio transmission between the mobile units and the base stations. Even though these systems are analog, adapters are available for FAX and modem transmission from mobile units. Many systems are being upgraded to digital modulation techniques using TDMA.

Each mobile unit is assigned a ten-digit number, identical in format to any other telephone number. Callers to the mobile unit will dial the number for the desired mobile unit. The mobile user will dial seven or ten digits with a "0" or a "1" prefix, where applicable, as if calling from a fixed telephone.

A mobile unit is called by transmitting its number over the setup channel. When the mobile recognizes its number, it quickly seizes the strongest setup channel and transmits an acknowledgment response. The cell site then uses the seized setup channel to transmit the voice channel assignment to the mobile. The mobile and cell site switch to the voice-channel radio frequency and the voice channel is used for ringing, off-hook, and subsequent conversations. The sequence is similar when the mobile user originates the call. It begins with the mobile control unit seizing a setup channel when the mobile unit goes off-hook. Then the voice-channel selection, signaling and conversation occur in the same way.

During the call, the system at the serving cell site examines the signal strength once every few seconds. If the signal level becomes too low, the MTSO looks for a cell site closer to the mobile unit to handle the call, based on the location and direction of travel from the serving cell site. The actual "handoff" from one cell to the next occurs so rapidly that the user normally cannot tell it has occurred.

All metropolitan areas in North America have full cellular coverage by two competing cellular telephone companies: a "wireline" operator (owned by the local telephone company) and a "nonwireline" operator (the independent radiotelephone common carrier).

Cellular radio is ideal in urban areas where, through frequency reuse, many simultaneous telephone conversations can be accommodated. This terrestrial system, however, is not as well suited for very remote or rural areas, where potential users are sparsely distributed.

As a result, a marriage of satellite communications and cellular phones is moving closer to reality. Consortiums of several well-known communication systems companies and manufacturers are developing global satellite communication systems. One in particular, called Iridium, will allow customers to call or be called anywhere on earth using hand-held wireless telephones. The system will relay telephone calls through a network of 66 low Earth orbiting satellites (LEOS) at an altitude of 780 kilometers (420 nautical miles). With the satellites orbiting more closely to the Earth's surface, the system can operate with lower power and will have lower round-trip delay than geosynchronous satellites.

The system will have six orbital planes, each with 11 operational satellites and one spare. As shown in Figure 16-3, when the satellite receives the signal, routing will be handled on Earth via a gateway station. The call can then be transferred from satellite to satellite and passed down to Earth. Communication links between the subscriber unit and the satellite will use the L-band (1,500 – 1,700 MHz) while inter-satellite links and gateways will use the Ka-band (17/30 GHz). Many Iridium telephones will be dualmode, allowing subscribers to interconnect with terrestrial cellular networks when available and compatible, and the constellation of satellites if cellular services are not accessible. The Iridium system is expected to be operational in 1998.

Figure 16-3   **The Iridium Network**

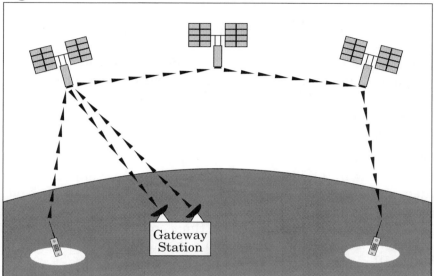

During the early 1990s, digital cellular technologies were introduced which offer the potential to initially triple the number of available switching channels to overcome the current shortage of cellular channels in large cities. This system is based on the time division multiple access (TDMA) standard (see Section 10) for North America.

With this change will come the opportunity to broaden the market and introduce new services. A new generation of low-cost, compact, digital cordless telephones for pedestrian and residential use will create a personal communications services (PCS) network. These pocket telephones will give immediate access to people and information without the geographical and physical constraints of traditional wired networks.

16.1 Personal Communications Services

Personal Communications Services (PCS) is expected to be the telecommunications industry's most significant growth area in the next decade. When PCS is fully operational, everyone could have a personal directory number that could be used to locate that individual anywhere in the world.

This concept requires the network instead of the caller to keep track of the individuals to be called. This represents a fundamental change from the way traditional networks have evolved. People usually have several directory numbers, each uniquely associated with the equipment in their homes, offices, cars, and cottages. As a result, people do not really call other people, they call places.

The market for wireless equipment, such as cordless phones, pagers and cellular phones has continued to expand. At the end of 1991, there were 7.6 million people in the United States using cellular phones. By 1996, the number of cellular users should top 30 million.

The PCS network will be based on a hybrid use of wireline technology and three classes of wireless technology:

- high-density, on-premises, low-power wireless systems
- high-speed, wide-coverage vehicular cellular systems and
- high-density, wide-coverage cellular micro systems

Low-power, in-building wireless systems, which provide mobility to users within a shopping center, airport or office building, are able to accommodate the high densities found in these environments. Low-power systems also use much smaller cellular structures than the vehicular cellular systems to provide coverage in high-density pedestrian areas (see Figure 16-4).

Cellular radio systems are designed for users who spend most of their time in a car traveling within a city. The third class of

Figure 16-4   **Low Power Wireless Network**

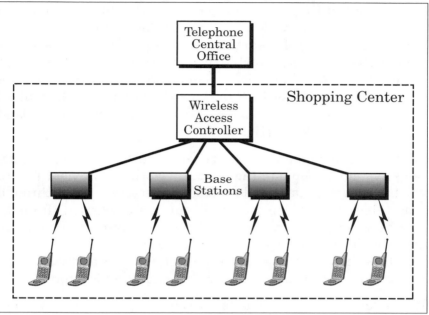

wireless technology, cellular micro system is targeted at those users who require mobility both within the office and on the road. This means the same handset can be used in the office and anywhere in the cellular network. As shown in Figure 16-5, when the user's handset enters a new area it automatically registers itself with the network's home location register (HLR), via the nearest base station and the controller. This database locates and tracks users as they roam throughout the cellular network. The wireless access controller communicates with the HLR over common channel signaling links.

The demand for PCS will be on a global basis but the regulations that govern the allocation of radio spectrum are determined country by country and are at varying stages of development. The first multinational standard for digital cordless (low-power wireless) telecommunications is the adoption of the CT2 common air interface (CAI) standard. The CT2 CAI standard ensures that equipment from different manufacturers is compatible, so that the same personal portable terminal can be used at home, at work and in public places. Spectrum bands that have potential for early development of wireless services include the 864-MHz to 868-MHz band allocated in Europe and some Pacific Rim countries. Canada has made 120 MHz of spectrum available for PCS in the 2-GHz band. In the United States, the FCC has allocated spectrum around the 2-GHz band as well as in the 900 MHz band.

With the development of low-power wireless telecommunications systems, people in the business world in the near future will be reachable via a single personal directory number through a portable, pocket-size terminal that can be carried throughout the work place, in transit, at home, or pursuing a leisure activity. The concept of a global PCS utilizing a network of satellites extends the reach of terrestrial PCS and will make personal communication networks virtually ubiquitous.

Whatever the future holds for wireless systems, such as digital cellular phones and PCS, there is no question that they have the potential to radically transform the telecommunications industry.

Figure 16-5   **Cellular Micro Systems**

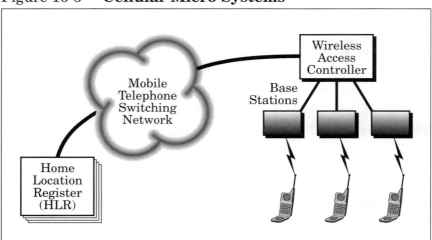

Conclusion

Cellular radio has proven to be one of the fastest-growing technologies in the world. The first system was tested in Chicago in 1977 and placed into full commercial service in 1983. In order to stimulate competition the FCC decided that two licenses would be issued in each area. Metropolitan areas are the best markets for the present terrestrial systems. The systems are designed to grow gradually to serve more users as demand increases. The larger markets have led to lower prices for customers.

The introduction of personal communications services (PCS) using pocket-size digital cordless telephones will open up a huge market. This will create direct competition between local telephone companies and cable TV companies for a share of this market. This development will enable people to be reached anywhere in the wireline, wireless and cellular networks. Subscribers will have a personal identification (ID) number, instead of a location-dependent telephone number. This ID number will enable callers to reach subscribers at a single number, regardless of whether they are at home, in the office or traveling.

This technological vision is moving closer to reality. The global PCS network, based on low Earth orbiting satellites (LEOS) at altitudes of about 1,000 km, and medium Earth orbiting satellites (MEOS) at altitudes of about 10,000 km, forming a chain around the globe, is anticipated to begin commercial service in 1998 and 1999. Geostationary satellites have a major advantage because continuous global coverage can be provided with only three satellites. However, the drawbacks include the round trip delay and the high power required. The LEOS minimize the transmit power and the delay but require a large number of satellites to provide global coverage. In additional, the speed of the LEOS relative to the Earth (about 7.4 km/s for Iridium) requires frequent handoffs. Despite the technical and economic considerations, the global PCS network will soon be a major component of our telecommunication network.

Review Questions for Section 16

1. What is the size of a typical cell in the present cellular network?

2. Explain how growth is handled in a cellular network.

3. What does MTSO stand for?

4. What radio frequency bands are assigned to a cellular system?

5. How many FM channels can be accommodated within those frequency bands?

6. Briefly describe the purpose of a "handoff."

7. How many satellites are required to provide global coverage on the Iridium network?

8. Define personal communications services (PCS).

9. Describe the three classes of technology in a PCS network.

10. What is the CT2 CAI standard?

17 Lightwave Systems

17.1 A Brief History

Just over 100 years ago, Alexander Graham Bell transmitted a telephone signal over a distance of 200 meters using a beam of sunlight as the carrier. That historic event involving the "photophone" marked the first demonstration of the basic principles of optical communications as it is practiced today. The photophone did not reach commercial fruition, however, due to the lack of a reliable, intense light source and a dependable, low-loss transmission medium.

As a result of the development of the laser, as a light source, and the optical fiber as a transmission medium, optical fiber transmission systems emerged in the late 1970s as a major innovation in the field of telecommunications.

Although the possibility of guiding light had been demonstrated in 1854, it was not until 1910 that a practical lightguide was envisaged. This was a hollow tube with a highly reflective metal coating on its inner surface, capable of guiding a wide range of electromagnetic waves, including those of visible light. However, the high signal loss of these devices, especially where they curved to change direction, made their use impractical. In the 1930s, experiments on another lightguide began. This time the guide consisted of simple filaments of glass fiber packed into bundles. By the 1950s, these glass fiber bundles were being used as light conduits for punched-card readers.

The development of optical fiber for use in telecommunications did not begin until the mid-1960s. It was initiated in 1966 by the publication of a paper by Dr. C.K. Kao (ITT England), in which he stated that pure optical fiber was theoretically capable of guiding a light signal with very little loss.

By 1970, Corning Glass Works in the United States was able to produce a fiber of sufficient purity for use in telecommunications. To make the fiber, Corning used a method of synthesizing silica glass. The raw materials were vaporized and deposited inside a length of quartz glass tubing, which was then collapsed into a rod and drawn into a fiber.

The technology for the other two key elements of a fiber optic system, the transmitter (emitter or light source) and receiver (detector), had been developed independently during the previous ten years.

In 1960, a possible candidate for the light source appeared, the ruby rod laser. This demonstration was followed in 1961 by the gas laser which could emit a continuous beam of light. By 1962, the expansion of solid-state technology had culminated in the semiconductor injection laser diode. This laser was very small, about the size of a grain of sand, and could turn an electrical signal into light while being modulated at extremely high speeds (see Figure 17-1).

Figure 17-1   **Injection Laser Diode**

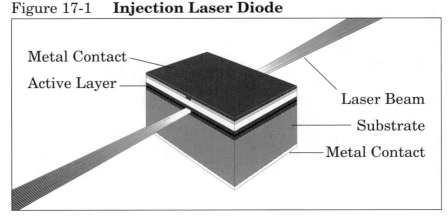

Lasers are not the only light sources for use in optical communications. The light emitting diode (LED) can also transform an electrical signal into light energy, although its light output is not as bright.

The final key element was the detector. This was a product of the silicon semiconductor industry. The photodiode was developed in the 1960s to detect light signals and turn them back into electrical information (see Figure 17-2).

The combination of these two unrelated technologies, semiconductor technology and optical waveguide technology, resulted in a transmission system that had many inherent advantages over conventional twisted pair systems. The lightwave transmission system was born.

In the late 1970s, a few telephone companies in the United States and Canada began to field trial experimental fiber optic transmission systems. One of the first was turned up in March 1979 by the British Columbia Telephone Company in Canada. At the time, this 7.4-km system was the longest optical telephone link and carried interoffice traffic at the 45-Mb/s rate without the aid of a repeater (see Figure 17-3).

Figure 17-2   **Side-Illuminated PIN Type Photodiode**

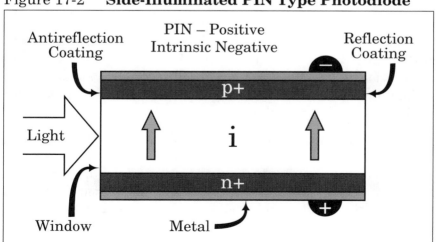

Figure 17-3   BC TEL Fiber Optic Field Trial

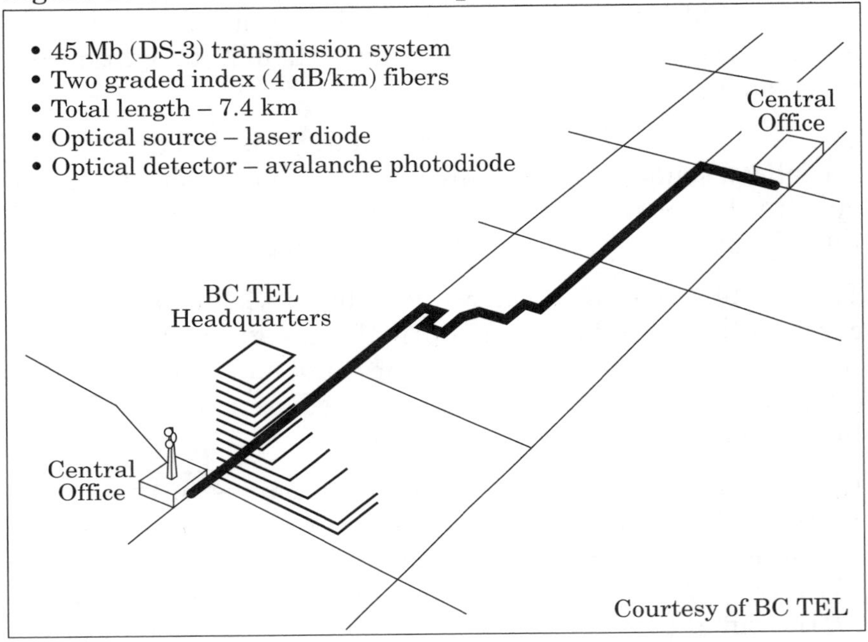

- 45 Mb (DS-3) transmission system
- Two graded index (4 dB/km) fibers
- Total length – 7.4 km
- Optical source – laser diode
- Optical detector – avalanche photodiode

Central
Office

BC TEL
Headquarters

Central
Office

Courtesy of BC TEL

These field trials demonstrated that lightwave systems were technically and operationally feasible for application in the telephone network.

The first operational systems began to appear in 1981. These systems were primarily designed for interoffice trunking and microwave radio entrance links. However, in 1983, a new generation of lightwave systems based on singlemode fiber technology operating at long wavelengths, became available for high-capacity, long-haul applications.

This development, along with the divestiture of AT&T in 1984, was one of the major reasons for the dramatic growth of lightwave networks. In addition, the deregulation of the long distance telephone industry in the United States created opportunities for other telephone carriers to establish networks of their own.

MCI, the second largest long distance carrier in the United States, and Sprint use fiber optic cable in their long distance networks almost exclusively.

AT&T, in an effort to modernize its networks, has scrapped billions of dollars worth of outdated equipment and replaced it with lightwave equipment and cable. A number of regional lightwave networks have also been developed. As shown in Figure 17-4, all major cities in the United States are now linked in a national lightwave network.

In Canada, Stentor has constructed a coast-to-coast lightwave system. Other common carriers are planning to built similar lightwave networks across the country.

New applications are emerging as fiber technology matures. One important application is in undersea cables because of the need for high-capacity transmission over long distances and between continents.

The introduction of singlemode long-wavelength technology has stimulated the development and growth of undersea systems. Lightwave undersea systems are currently operating across both the Atlantic and Pacific oceans.

Figure 17-4   **U.S. National and Regional Lightwave Networks**

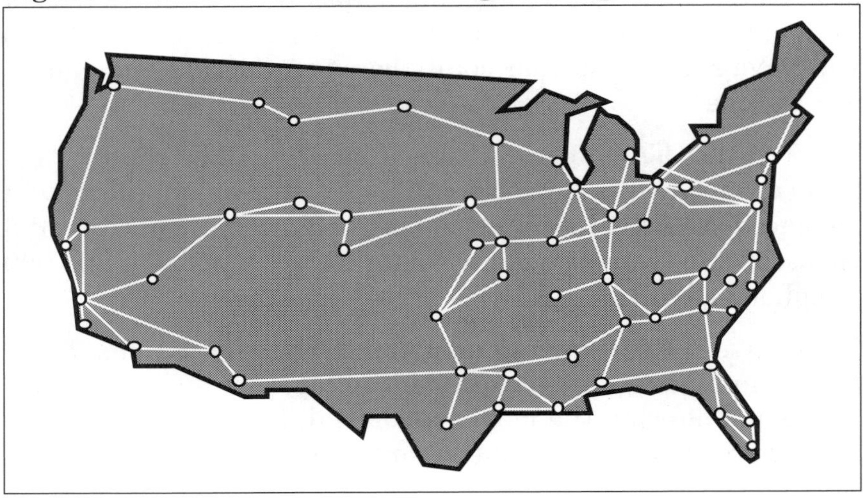

TAT-8, installed in 1988 to provide service to the United Kingdom and France, was the world's first transatlantic cable to use lightwave technology. TAT-9, installed in 1991, is capable of carrying 80,000 simultaneous telephone calls, about twice the capacity of TAT-8. HAW-4 and TPC-3 in the Pacific Ocean were put in service in 1989.

The North Pacific Cable (NPC) linking Pacific City, Oregon, with Miura, Japan, with a spur to Alaska, was turned up for service in 1990.

TPC-4 is in service linking the United States and Japan directly instead of going via Hawaii. TPC-5 is scheduled for service in 1996 and will link Japan, Guam and Hawaii with the United States. A 7,400 km cable (Pac Rim West) connects Australia and Guam. An 8,600 km cable (Pac Rim East) connects Australia and Hawaii.

This is all part of a plan to join Malaysia, Singapore, Brunei and the Philippines in the west, with Hong Kong, South Korea and Japan in the north, Australia in the south and Guam and Hawaii in the center with the United States by the latest in undersea lightwave technology (see Table 17-1).

Teleglobe Canada has completed the construction and laying of CANTAT 3, a 7,500 km cable across the Atlantic. With a total capacity equivalent to 60,480 voice circuits, CANTAT 3 will have six terminating points between Canada and Northern Europe.

The TAT-12 / TAT-13 transatlantic cables will be the first cables equipped with optical amplifiers. Optical amplifiers will be placed at 45 km intervals on both cables. Transmitting at 5 Gb/s, the two systems will have a total capacity of over 300,000 telephone circuits.

Lightwave systems have demonstrated their operational capabilities and have gained worldwide acceptance. They are now capable of offering the most economical, flexible and expandable approach to long-haul transmission.

Table 17-1   **Major Lightwave Undersea Systems**

System	In-service Date	Terminating Points
Transatlantic Systems		
TAT-8	1988	U.S., U.K., France
TAT-9	1991	U.S., U.K., France, Spain, Canada
TAT-10	1992	Canada, U.K., Europe
TAT-11	1993	U.S., U.K., France
TAT-12	1995	U.S., U.K., France
TAT-13	1996	U.S., France
CANTAT 3	1995	Canada, Iceland, Faeroe Islands, Germany, Demark, U.K.
PTAT-1	1989	U.S., U.K., Bermuda, Ireland
PTAT-2	1992	U.S., U.K.
TAV-1 (TV)	1990	U.S., Europe
Transpacific Systems		
TPC-3/Haw 4	1989	Japan, Guam, Hawaii, U.S.
TPC-4	1992	Japan, U.S.
TPC-5	1996	Japan, Guam, Hawaii, U.S.
NPC	1990	U.S. Alaska, Japan
Transindian Systems		
SE-ME-WE 2	1994	Marseilles, Singapore
SE-ME-WE 3	1999	Europe, Middle East, Asia
Pacific Basin Systems		
G-P-T	1989	Guam, Phillippines, Taiwan
H-J-K	1990	Hong Kong, Japan, Korea
Tasman-2	1991	Australia, New Zealand
PacRim East	1993	Australia, Guam, U.S.
PacRim West	1994	Australia, Guam, S.E. Asia
Europe to Asia System		
Flag	1997	U.K., Middle East, Africa, Asia

17.2   Optical Fiber Cable

The design of an optical cable requires attention to several additional physical parameters over those required in a conventional twisted copper pair cable. The transmission properties of a fiber are critically dependent on the stresses it encounters, because any stress-induced density fluctuations in the glass can both scatter light out of the fiber and can change its propagation characteristics. Thus, cables are designed to minimize fiber stresses during construction and installation and over a wide range of environmental temperatures.

The first step in cable construction is to protect fibers individually before they are jacketed together. The two techniques are known as loose and tight-buffer construction.

In the loose-buffer construction, shown in Figure 17-5, the fiber lies with a slight helical wind within a soft polymer tube. This tube provides cushioning and also isolates the fiber from axial

Figure 17-5   **Loose-Buffer Tube**

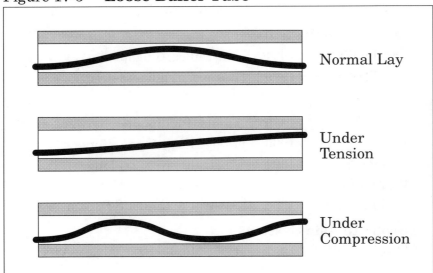

Normal Lay

Under
Tension

Under
Compression

strains, both tension and compression. When the cable is under tension, the free-floating fiber will tend to straighten its helical lay within the tube to avoid any stress. While under compression, the fiber will tend to coil up in the tube. Thus, the loose buffer construction isolates the fiber from installation or environmental stresses.

In tight-buffer construction, shown in Figure 17-6, a soft polymer layer is extruded directly in contact with the fiber, which serves to cushion it from external strains. One of the advantages of a tight buffer is its accessibility for attaching a connector. The loose-buffer tube makes the fiber more difficult to attach a connector. However, the loose buffer adds less optical attenuation to the fiber, and can be built to withstand wide temperature specifications.

After the individual fibers have been protected, they can be stranded into the final cable structure. The basic ingredient of most optical cable structures, similar to that of conventional cables, is plastic, polyethylene and polyvinylchloride. A typical loose-buffer tube cable is shown in Figure 17-7.

Figure 17-6   **Tight-Buffer Cable**

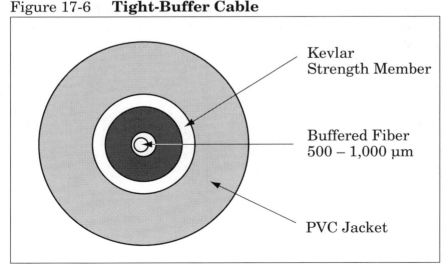

Kevlar
Strength Member

Buffered Fiber
500 – 1,000 µm

PVC Jacket

Figure 17-7   **Loose-Buffer Tube Cable**

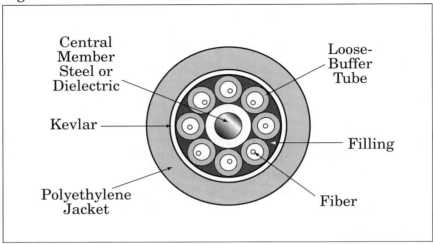

To resist buckling and add tensile strength, many cable designs contain a stiff central element, such as steel or epoxy/fiberglass composite. This also serves as a stabilizing element around which the fibers can be helically wound. Usually, the central strength member is cushioned to minimize stresses on the fiber under crush conditions.

Similar to conventional cables, some optical cables are constructed with metallic sheaths and armor. This provides crush resistance, tensile strength and a seal against water.

17.3   Optical Fiber Transmission Parameters

The four primary parameters which specify the characteristics of an optical fiber are numerical aperture, attenuation, dispersion and bandwidth.

• Numerical Aperture

Numerical aperture (NA) is a measure of a multimode fiber's light acceptance angle.   Rays of light striking the end of the fiber are   efficiently transmitted through the fiber only if their angle of incidence is within a specified number of degrees of the fiber's axis. Typical values for the NA of a telecommunications type of fiber would be from 0.16 to 0.25, corresponding to maximum acceptance angles of 9° and 14° with respect to the axis (see Figure 17-8).

Figure 17-8   **Numerical Aperture**

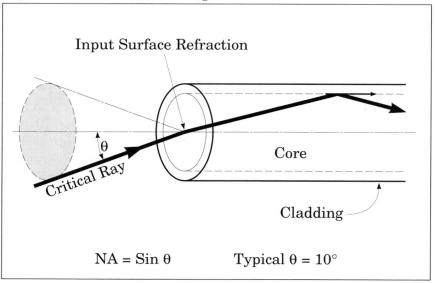

• Attenuation

An optical fiber consists of a transparent core, confining most of the light, surrounded by a transparent depressed index cladding that has a slightly lower index of refraction. Today's best fibers are made of silica. The core is usually doped with germanium and possibly phosphorous pentoxide to enhance its refractive index. The cladding may be pure silica or doped with boron oxide at short wavelengths or fluoride at long wavelengths to depress its refractive index.

Silica optical fibers are the most transparent materials manufactured, so an optical signal with a power of only a few milliwatts can be detected after it has traveled dozens of kilometers. However, a fiber does have losses, the signal traveling though it gradually becomes faint, and eventually a repeater is needed to boost and regenerate the signal. The installation and maintenance of these repeaters can be a major expense in a lightwave system. Longer repeater spacings can be attained if the fiber's attenuation is reduced.

In the 1970s researchers found that attenuation depended on wavelength. A typical singlemode fiber has three windows as shown in Figure 17-9. The first window at 850 nanometers (nm) has an attenuation of approximately 2.5 dB per kilometer of fiber. Some of the light, however, is absorbed by hydroxyl (OH) ions trapped during processing. This "water impurity" produces absorption peaks at 1250 nm and 1390 nm. Though processing can reduce the "water peaks," it is simpler and less costly to avoid the problem entirely. At 1300 nm the attenuation is only 0.5 dB/km. However, on the other side of the water peak at 1390 nm, the fiber's attenuation decreases again, reaching its greatest transparency of 0.25 dB/km around 1550 nm. As a result, 1300 nm and 1550 nm wavelengths have been chosen for long-haul, high-bit-rate lightwave systems. Optical fiber loops presently use the 1300 nm window almost exclusively.

Figure 17-9   **Optical Fiber Attenuation as a Function of Wavelength**

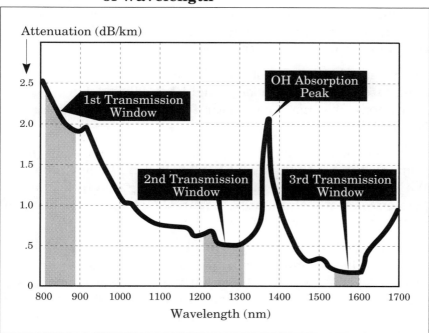

• Dispersion

Attenuation is not the only characteristic of silica fiber that affects repeater spacing. Multimode distortion, which is also called modal dispersion, can cause pulse spreading in multimode fibers. The earliest fibers, known as "step index" types, caused almost complete internal reflection of each of the modes in the core so that most of the energy was contained and propagated along the fiber. In multimode fiber not all the light rays take the same path down the fiber core. Some travel straight down the core, but others are reflected off the interface between the core and the cladding, or are refracted within the core, following a zigzag path. The rays (or modes) taking the zigzag path must travel a longer distance to the receiver and are therefore somewhat delayed.

Improvements in multimode fibers included a type known as graded index. To reduce this delay difference, the interface between the core and the cladding is carefully doped to create a gradient in the refractive index that is parabolic, so the rays travel faster near the interface.

The most effective way of reducing multimode distortion is to eliminate all but one mode of light traveling down the fiber. Multimode fiber has a core with a diameter of 50 microns (see Figure 17-10).

By reducing the core diameter to 8 microns only one mode remains, the one that effectively travels straight down the core. With singlemode fiber the pulse spreading due to modal dispersion is mostly eliminated. Figure 17-11 shows the paths of light in the three types of fiber.

Singlemode fiber also suffers from chromatic (or material) dispersion. The chromatic dispersion arising from the fiber's material properties causes the slightly differing wavelengths of emitted light to travel through the fiber at different speeds. As a result, an optical pulse will broaden and spread out as it travels through the fiber. This pulse broadening increases with

Figure 17-10 **Singlemode & Multimode Fiber**

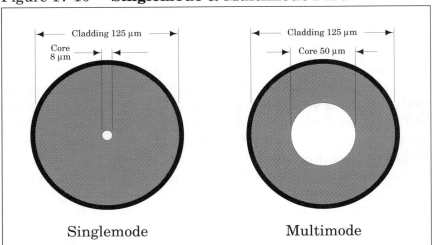

Figure 17-11   **Light Paths**

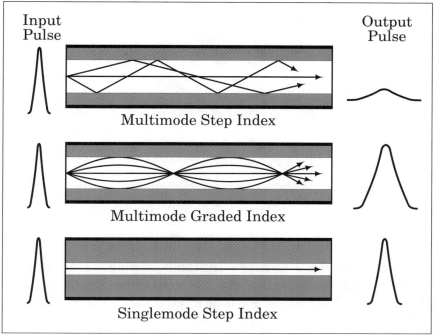

the length of the fiber between the repeaters. If the chromatic dispersion becomes too great, each pulse becomes broad enough to interfere with its neighboring pulses, which increases the bit error rate. Therefore chromatic dispersion forces a trade-off between the spacing of repeaters and the effective bit rate.

Because of the intrinsic properties of silica, the amount of chromatic dispersion varies with the wavelength of light. At any particular wavelength, dispersion is measured as the amount of delay in picoseconds per kilometer of fiber per nanometer change in the wavelength. At 850 nanometers the chromatic dispersion is typically 90 ps/km-nm. The amount of dispersion decreases rapidly with increasing wavelength, until it reaches zero at 1313 nm, after which it increases again (see Figure 17-12). Unfortunately, at 1550 nm, where attenuation is at a minimum, the chromatic dispersion reaches about 18 ps/km-nm.

Figure 17-12 **Dispersion Wavelength Dependency**

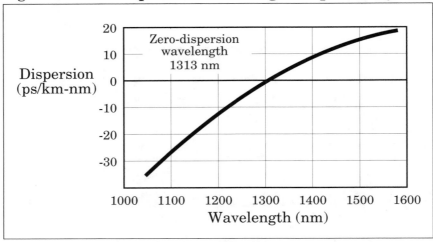

However, in 1985, a new singlemode dispersion-shifted fiber was introduced, in which the zero dispersion point was shifted from 1313 to 1550 nm. Systems using 1550 nm lasers and detectors over dispersion-shifted fiber have the best of both worlds, the lowest attenuation and minimum dispersion. However, dispersion-shifted fiber is more expensive.

• Bandwidth

The capacity of a communications system is directly proportional to its bandwidth: the wider the bandwidth, the greater its information-carrying capacity. For comparison purposes, it is common to express the bandwidth of a system as a percentage of its carrier frequency. For example, a microwave radio system operating at 6 GHz with a bandwidth equal to 10% of its carrier frequency would have a bandwidth equal to 600 MHz.

Light frequencies used in lightwave systems are between 10^{14} and 10^{15} Hz (100,000 to 1,000,000 GHz). Therefore the theoretical capacity of a lightwave system is 10% of 1,000,000 GHz or a staggering 100,000 GHz!

17.4   Fiber Connectors

In the 1970s interconnection problems were cited as major barriers to optical fiber technology. The first systems did experience a number of outages as a result of connector problems. However, recent years have seen dramatic changes in the products and techniques for interconnection.

The basic purpose of a fiber optic connector is the nonpermanent connection of an optical fiber to another optical fiber or to an active device such as an emitter or detector. A splice, on the other hand, is a nonseparable junction of two fibers. The insertion loss of the connector primarily reflects the difficulty of maintaining precise tolerance in a detachable connector. Additionally, the profound effect of even a small amount of dirt makes this component particularly critical. In addition to having an acceptably low insertion loss, a connector must also show good repeatability of this loss over many connect / disconnect cycles.

The connector designs that have emerged over the past few years are available in many varieties, such as the SMA, the ST®, the FC, the SC and the biconic.

The SMA-style connector (shown in Figure 17-13) provides noncontacting connections. The exposed fiber must be properly polished to create a low-loss connection, and since the connector is threaded, it must be properly tightened. Losses are usually in the range of one dB. The FC connector has largely replaced the SMA where threaded connectors are required.

The ST® style (ST is a registered trademark of AT&T) (shown in Figure 17-13) uses a twist-on locking mechanism making it easier to "feel" when it is properly seated. In addition, the connector is designed as a contacting type, decreasing the maximum loss to less than one dB. The ferrules can be constructed from stainless steel or ceramic.

Figure 17-13   **SMA and ST® Style Connectors**

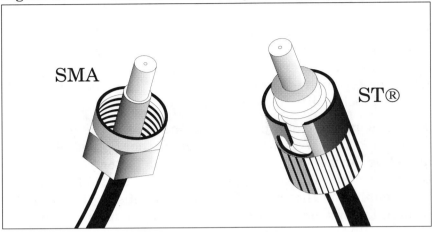

SMA

ST®

Figure 17-14   **Biconic Type Connector**

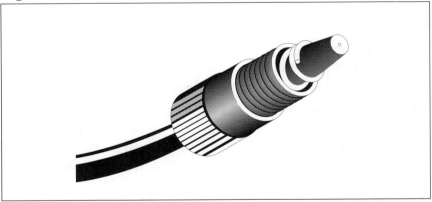

The biconic connector (see Figure 17-14) uses a contacting cone-shaped ferrule. Like the SMA and FC, the biconic is threaded. They require a precision polishing operation, making their installation a bit more cumbersome than the SMA, FC or ST® devices. As a result, the SC connector has become very popular because of its push-pull operation and is used by many telcos.

Installation kits are available which include a variety of tools for cutting the cable and fastening the connectors to the fiber. Quick-cure epoxies are supplied to secure the exposed fiber in the ferrule before polishing.

17.5   Optical Emitters and Detectors

Optical emitters and detectors are the transmitters and receivers of lightwave systems. The optical properties of these components can determine the ultimate performance of the lightwave system. One type of emitter is the light emitting diode (LED). These devices have a relatively linear light output versus electrical input transfer function, and are therefore useful for the transmission of both analog and digital signals. LEDs, however, do have some performance limitations. The output power is relatively low, on the order of −15 dBm. Additionally, the output spectrum of these devices is very broad, on the order of 100 nm. Many low-capacity and short-haul systems use LEDs for emitters.

Laser diodes, on the other hand, have significantly more light output available, up to +10 dBm. Laser diodes also have relatively narrower optical spectra than do LEDs, around 10 nm. The light output versus current transfer curve of a laser is very nonlinear, so these devices are usually best suited for the transmission of digital signals. This type of device is known as a Fabry-Perot (FP) type. With these types of devices, transmission of 565 Mb/s over 40 km without repeaters is fairly standard today, and many systems of this type are in operation for long distance telephone transmission.

A newer laser, the distributed feedback (DFB) laser, behaves much more like a single frequency source with a spectral width of 0.5 nm. Figure 17-15 illustrates how the spectral shape of the emitted light from the two lasers is determined. The upper part of the illustration shows cross sections of the FP and DFB laser. In the FP laser, the active layer forms a cavity resonator having the shape of a thin and narrow stripe with cleaved mirror facets at both ends, enabling standing waves of light to form in it. Therefore, the spectral shape of the emitted light is determined by the frequencies of standing waves that exist within the gain bandwidth of the active layer.

As shown on the lower portion of this figure, the spectrum of the FP laser is composed of a number of spectral lines, resulting in an effective spectral width of several nanometers. On the other hand, in the DFB laser there is a corrugation structure close to the active layer. In addition, the light-emitting facet is processed with antireflective coating to reduce mirror reflection so that the FP mode of oscillation is suppressed. Because the corrugation provides distributed feedback for light in a very narrow spectral range, the lasing occurs with a pure single frequency. The lasing spectrum is thus a single line having a very narrow spectral width even when the diode is modulated with a high-speed signal. Because of this, the degradation of the transmitted signal due to fiber chromatic dispersion is dramatically reduced, giving a corresponding capacity increase.

Detectors utilized by high-capacity systems are usually one of two types; a PIN diode, with a window opened into the diode to allow the entry of light (see Figure 17-2), and the avalanche photodiode (APD). The PIN diode produces, at best, one electron per photon of incident light. The PIN photodiode must therefore be followed by a very low noise amplifier to obtain good receiver sensitivity. Higher sensitivity has been obtained with the APD devices. These devices are diodes that are reverse-biased near the avalanche region. When a photon

Figure 17-15   **Fabry-Perot and DFB Laser Diodes**

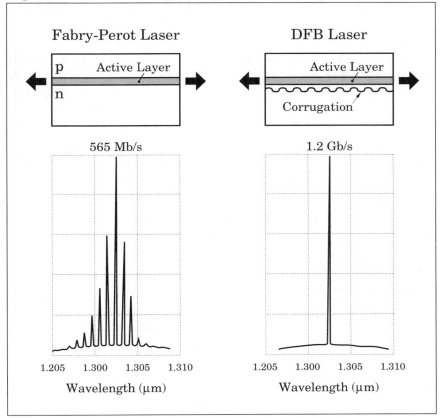

strikes the detector, it generates a hole-electron pair. The high electric field causes the charges to be rapidly swept to the diode connections. The sweeping is so rapid that additional electrons and holes are generated in the process. Thus, photomultiplication occurs. A germanium (Ge) APD generally runs with a multiplication factor of 10 at a 565 Mb/s data rate. Since the dark or leakage current and the avalanche phenomenon are somewhat temperature sensitive, the detector usually must be cooled to maintain good sensitivity, usually to no greater than 20°C or 25°C.

The germanium APD is currently a very common detector. It suffers from excessive dark current (thermally induced) and excess multiplication noise, and also has reduced frequency response at 1550 nm compared to 1310 nm. A newer alternative is the ternary or indium gallium arsenide (InGaAs) APD, which has lower dark current and lower multiplication noise, and provides a higher useful gain than the Ge device at high data rates. The gain of the APD helps reduce the effect of the noise in the following electrical amplifier. The InGaAs APD has an improved gain-bandwidth product compared to other devices at a wavelength of 1550 nm (where typical singlemode fiber has the lowest attenuation) due to a smaller multiplication region which has a high carrier concentration.

As the demand for broadband services increases, such as video-conferencing, imaging, and pay-per-view television, the telecommunication companies will be upgrading their networks to deliver services at increasingly higher speeds. Today lightwave systems operate at 2.4 Gb/s. This means a single fiber can support 32,000 voice channels or 48 studio-quality TV broadcasts. By the year 2000, lightwave systems will have to support speeds beyond 10 Gb/s.

The world's leading designers and suppliers of lightwave equipment are developing the technologies capable of breaking through the 10 Gb/s bandwidth barrier and establishing new fiber-span distance records. These devices include new lasers, detectors and other electronic and optoelectronic products. One new device is a semiconductor phase modulator which minimizes chirping at high frequencies by eliminating the need to turn lasers on and off. Chirping is the small changes in wavelength that occur when a laser is turned on and off at high speeds causing transmission errors. The modulator acts as a shutter that rapidly opens and closes, producing very short light pulses with minimal frequency distortion at 10 Gb/s.

As the year 2000 approaches and the information superhighway becomes a reality, lightwave systems will be capable of carrying all the interactive services that the public will need.

17.6   The Evolution of Lightwave Systems

The signal characteristics of a lightwave system are dramatically different from those of a conventional transmission system. When dealing with ultrahigh-frequency electromagnetic waves, such as light, it is common to use units of wavelength rather than frequency (cycles per second or Hertz), as is the case with conventional systems. The frequencies used for lightwave systems extend from approximately 10^{14} to 10^{15} Hz (infrared to ultraviolet). Most fiber systems operating today have a wavelength of 1310 or 1550 nanometers (nm). This wavelength is beyond the sensitivity of the human eye (which cuts off at about 750 nm) in the near-infrared portion of the spectrum.

The first operational systems back in 1981 operated at a wavelength of 850 nm with a bit rate of 45 Mb/s, providing a maximum capacity of 672 voice circuits over two fibers.

A simplified diagram of a typical system is shown in Figure 17-16. There, 1.544 Mb/s (DS-1) signals from a channel bank (24 voice circuits) are combined with 27 other DS-1 bit streams into a 45 Mb/s (DS-3) bit stream by an M13 multiplexer.

The resulting DS-3 stream is then converted into optical pulses by the laser diode operating at a wavelength of 850 nm. After transmission through the fiber cable, the optical pulses are detected by an avalanche photodiode and converted back to electrical signals, regenerated and then reconverted into a standard DS-3 signal for input to the M13 multiplexer and the associated channel banks.

The third-generation fiber optic systems operated at 1300 nm with a bit rate of 565 Mb/s providing 8,064 voice circuits per fiber pair (see Figure 17-17). The systems now installed are designed to be upgraded to 1.2 or 2.4 Gb/s. These systems could then carry four times as many calls simply by installing new

Figure 17-16   **Lightwave Transmission System**

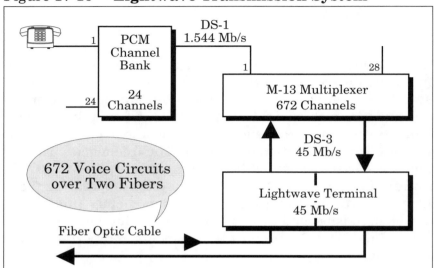

Figure 17-17   **565 Mb/s Lightwave Terminal**

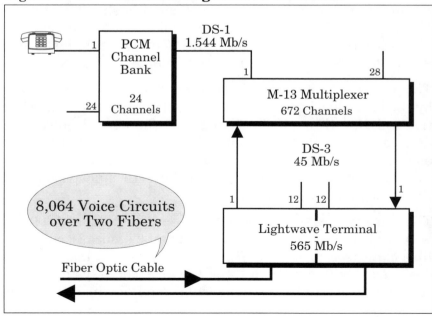

electronics in terminals and repeater stations along the routes. The fourth-generation systems (see Figure 17-18) operate at 1550 nm and have about half the fiber loss as compared to the 1300 nm systems. These systems have repeaters spaced up to 100 km apart, and a bit error rate of five errors per 10 billion bits. In other words, if you transmitted the text of four 30-volume sets of the Encyclopedia Britannica over a distance of 100 km in one second, the only error would be that one letter might be capitalized instead of lowercase!

The fifth generation will probably feature coherent detection. Coherent detection is the lightwave equivalent of a microwave radio receiver that heterodynes the received signal to an intermediate frequency (IF). The term coherent refers to the use of a laser as a local oscillator in the receiver. It should provide a 10 dB to 20 dB improvement in receiver sensitivity and thereby increase the repeater spacing.

Figure 17-18   The Four Generations of Lightwave Systems

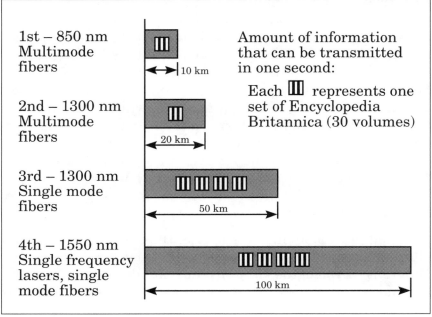

Obviously, there will be many other factors (for example, economic, standards and customer demand) that will play a significant role in determining the future direction of lightwave development.

17.7   System Design Considerations

High capacity, singlemode long-wavelength lightwave systems are now available from at least ten manufacturers. These systems are unaffected by rain, temperature or humidity, and are impervious to electromagnetic interference.   Typical bit error rate (BER) performance for lightwave systems is one error in 10^{11} bits.   Therefore, the performance of lightwave systems is substantially superior to that of terrestrial microwave, satellite or twisted pair transmission systems.

At optical line rates of 140 Mb/s or less, the system gain of the electronics and the attenuation of the particular cable design are all that is really required to determine whether or not an overall electronics and cable system will function properly at the required bit error rate. At bit rates of 565 Mb/s and above, however, several new parameters enter into the system span design calculations. These parameters, which determine the total dispersion, are the dispersion of the fiber cable, the spectral line width of the laser transmitter, and the maximum allowable dispersion of the receiver for a given bit error rate. At the higher speeds either insufficient system gain or excessive dispersion will limit the maximum span length that may be used in designing a lightwave network.

As noted previously in Section 17.3, the dominant dispersion effect in singlemode optical fiber cable is material dispersion. Different wavelengths of light propagate down the glass core at different velocities.   In an ideal cable, these wavelengths would all arrive at the detector at the same time. The dispersion coefficient of a fiber, in units of ps/nm-km, is an indicator of the extent that the fiber varies from the unattainable ideal. The dispersion coefficient of a fiber is a well-known function of the wavelength that is being transmitted down the fiber. This means that by picking a proper transmit wavelength, one may limit the cable dispersion effect to some degree. Dispersion is also a function of distance: the longer the fiber, the worse the dispersion.

It should be noted that cable manufacturers guarantee only certain specifications, and these specifications are usually very conservative. Therefore, to achieve a greater span length, consideration must be given to the design of the laser transmitter. Within the transmitter, two actions may be taken. Either the spectral line width of the transmitter may be reduced, or the transmit wavelength may be changed to correspond with the minimum dispersion points of the optical fiber cable.

Some manufacturers choose to use distributed feedback (DFB) lasers in their products. These lasers have very narrow spectral line widths, on the order of 0.5 nm. With these lasers, the dispersion of the fiber cable does not limit the span length since the transmitted pulse is so narrow from a wavelength standpoint. Attenuation limitations come into play before dispersion limitations under these conditions. The drawback to DFB lasers is their cost, which can be significantly higher than the more standard Fabry-Perot (FP)-type lasers.

Other manufacturers use the FP-type laser and screen each laser transmitter for transmit wavelength and spectral line width. Since the worst-case fiber dispersion is known as a function of wavelength, then the maximum allowable spectral line width for any transmit wavelength is also known. Manufacturers screen the transmitters for this required transmit wavelength / spectral line width relationship. With the current fiber specifications in the marketplace, this screening process allows good dispersion performance to fiber lengths of 45 km. At this distance, the span lengths of systems using either

DFB or FP lasers typically start to become limited due to attenuation.

Thus, for span lengths of 45 km or less, the system dispersion is typically not the limiting factor. Attenuation and system gain are the main elements to be considered during the span design process.

The supplier of the fiber cable must provide the required information to complete the attenuation calculations. In addition, the cable supplier must identify the amount of dispersion in ps/nm-km, over wavelengths between 800 and 1600 nm. The general specification for the fiber should also include such items as core diameter, numerical aperture, mode field diameter, cladding diameter, mode field concentricity error, refractive index profile, cutoff wavelength and temperature sensitivity (over a –50°C to +50°C) range.

For additional information on fiber cable specifications please refer to *"Subscriber Loop Signaling and Transmission Handbook: Digital"* by Whitham D. Reeve, published by IEEE Press in 1995.

The supplier of the electronics should clearly indicate the performance parameters of the system in order to develop a link loss budget. The budget should include, as a minimum, the following parameters for the equipment and cable:

- Transmit output power in dBm
- Initial cable attenuation (including connector and splices losses) in dB
- Receiver sensitivity in dBm
- Allowance for fiber aging variation in dB
- Allowance for maintenance splices in dB
- System margin in dB
- Power penalties in dB

A typical link loss budget and level diagram is shown in Figure 17-19. We have assumed for our design a minimum laser output of –3.5 dBm and a receiver sensitivity of –36 dBm.

Figure 17-19   **Typical Link Loss Budget and Level Diagram**

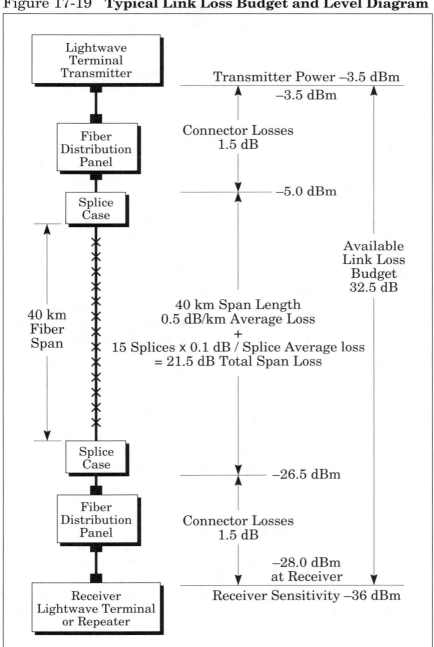

161

Using singlemode fiber with a loss of 0.5 dB/km and assuming a maximum of 15 splices at 0.1 dB per splice, a span of 40 km can be achieved with a comfortable operating margin of 8 dB.

An important consideration in the design of a lightwave system is the development of a strategy to expand the capacity of the system to meet future growth.

Today, long-haul singlemode fiber systems typically run 40 to 80 km between regenerators. They currently operate from 565 Mb/s up to 2.4 Mb/s in the 1550 nm window. As mentioned in Section 17.5, future systems will have to operate at 10 Gb/s in order to meet the demand for interactive broadband services.

One technique available to rapidly expand the capacity of singlemode fiber systems is to "stuff" more than one optical wavelength into a single fiber. This technique, called wavelength division multiplexing (WDM), permits a system to operate at both 1300 and 1550 nm wavelength simultaneously. However, WDM must utilize lasers that can emit wavelengths within very tight tolerances so they can be distinguished from each other when detected at the receiving end of the fiber.

The recent development of erbium doped fiber amplifiers (EDFA) is a major milestone in lightwave communications technology. Present lightwave systems use regenerative repeaters to compensate for signal attenuation and dispersion over long transmission distances. The optical signal is converted to an electrical signal, amplified by electronic circuits and then re-converted back to an optical signal. However, if the optical signal can be directly amplified, the repeater can be smaller and cheaper than existing regenerators.

Figure 17-20 shows the basic configuration of an erbium doped amplifier. The gain medium is a fiber doped with a very small amount of a rare earth ion, erbium. The optical signal is operated in the 1550 nm window. To induce gain in the doped fiber, the optical signal is optically pumped by a laser diode. The optical pump power is combined inside the fiber core by a wavelength division multiplexer. The pump laser operating at 1480

Figure 17-20 **Erbium Doped Fiber Amplifier**

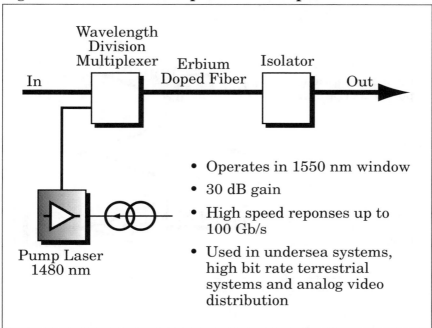

Wavelength
Division
Multiplexer   Erbium   Isolator
              Doped Fiber

In   Out

Pump Laser
  1480 nm

- Operates in 1550 nm window
- 30 dB gain
- High speed reponses up to 100 Gb/s
- Used in undersea systems, high bit rate terrestrial systems and analog video distribution

nm offers the best absorption band for the erbium ion. The optional optical isolator is used to avoid optical feedback and laser oscillation. A gain of 30 dB is possible.

As shown in Figure 17-21, an optical amplifier, such as the EDFA, can be used as a post amplifier increasing the output power of the transmitter, an in-line amplifier between two fiber spans, or a preamplifier in front of the receiver.

The first transatlantic cable utilizing erbium doped fiber amplifiers will be TAT 12 scheduled to enter service in 1995. The EDFA could revolutionize undersea communications, high bit rate terrestrial systems and analog video distribution systems. In fact a longer-term application for EDFAs might be the opportunity of exploiting the nonlinear propagation of optical signals called solitons. Solitons are optical signals propagating practically indefinitely without distortion.

Figure 17-21   **Applications of Erbium Doped Fiber Amplifiers**

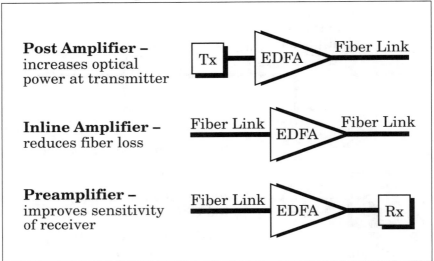

Post Amplifier – increases optical power at transmitter

Inline Amplifier – reduces fiber loss

Preamplifier – improves sensitivity of receiver

Optical amplifiers are the first step towards all-optical networks. While no one knows what new services will travel over the information superhighway of tomorrow, new users and new applications will bring many challenges. Network capacities will be stretched to the limit in order to deliver a range of applications with different performance and bandwidth requirements.

The solution may lie with all-optical networks, which carry information on beams of light from one end of the network to the other. Unlike today's fiber networks, which typically carry single wavelengths point-to-point across the network through intermediate stages of optoelectronic conversion, all-optical networks will multiplex, amplify, and route multiple wavelengths entirely in the optical domain without the need for conversion. These new networks will widen the information superhighway for such emerging services as remote medical imaging, remote business and banking transactions, Internet and Local Area Network (LAN) access.

17.8   SONET (Synchronous Optical Network)

As outlined in Section 6, the design of the North American digital network is based on hierarchical levels of DS-1 (1.544 Mb/s), DS-2 (6.312 Mb/s) and DS-3 (44.736 Mb/s). Since the introduction of digital technology in 1962, the growth in the digital network has been at the DS-1 rate.

With the introduction of high-capacity digital radio and lightwave systems, however, the DS-3 rate has also become a fast-growing portion of the network.

Since most network services are offered at rates lower than DS-3, there is a requirement to easily access and manipulate the lower-rate parts of the DS-3 signal. Unfortunately the signal format of the current DS-3 rate does not allow for efficient access and manipulation.

The DS-3 signal is the result of a multistep, partially synchronous and partially asynchronous multiplexing scheme (see Figure 17-22). The first step in this scheme involves synchronous byte interleaving of 8-bit words from each of 24 DS-0 (64 kb/s) bit streams to form a DS-1 signal, or digroup, of 1.544 Mb/s. Each DS-0 is timed from the same clock used to produce the DS-1 bit stream, but each DS-1 in the network may derive its clock from an independent crystal oscillator or other timing source.

The second step involves bit-by-bit interleaving of information from four DS-1 signals to form the DS-2, or 6.312 Mb/s, rate. But since the frequencies of the DS-1 signals are assumed not to be equal (that is, they are asynchronous), there must be some mechanism for bringing them to a common frequency before this bit interleaving can take place. This mechanism is called pulse stuffing, in which dummy information (stuffing bits) is added as necessary to each tributary DS-1 to bring it up to a common frequency prior to interleaving. Finally, in the third

Figure 17-22 **Synchronous Byte Interleaving of 8 Bit Words**

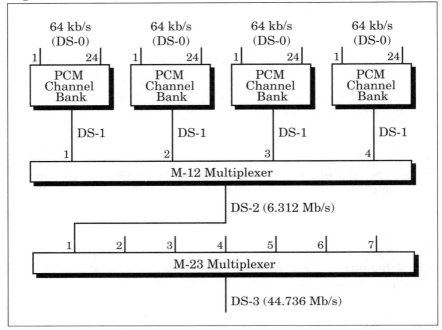

step of the process, seven DS-2 signals are combined by a similar method of pulse stuffing and bit interleaving to form the DS-3 signal.

Once the DS-3 level framing bits have been located, it is still not possible to say what DS-0, or even what DS-1, the following bits came from. The first bit may be from DS-1 number 1, the next from number 7, the next from number 10, the next a "blank" or stuffing pulse, and so forth. The only way to decode the DS-3 signal and locate the bits from a particular DS-0 or DS-1 is to reverse the process by which the DS-3 stream was created, demultiplexing one step at a time.

The current DS-3 signal format has been supplemented by two newer standards based on synchronous multiplexing. The first standard, known as SYNTRAN (for synchronous transmission), overcomes many of the problems in dealing with the current signal. SYNTRAN results in a simplified multiplex arrangement and reduced switch termination costs, promotes new and more flexible and efficient network architectures and offers significant operational advantages associated with fully synchronous digital transmission.

In today's network, if a DS-3 rate (44.736 Mb/s) signal passes through a repeater building (say, from west to east) and a few DS-1s (1.544 Mb/s) have to be dropped off, the entire DS-3 signal must be demultiplexed into 28 DS-1s. After the few DS-1s are dropped to local channel banks or a digital switch, the remaining DS-1s and any DS-1s to be added to the eastbound signal are then remultiplexed into another DS-3. If say, only three DS-1s are added and dropped, then roughly 25/28 of the circuitry of both multiplexers is wasted.

With SYNTRAN, because it is possible to identify the DS-1s directly, up to 28 DS-1 signals can be dropped or added from a DS-3 signal and the expensive back-to-back multiplexers can be eliminated. In addition, you can remotely select any of the available 672 DS-0 (64 kb/s) channels to define the contents of each DS-1 signal. This capability facilitates the grooming and consolidation of switched or nonswitched circuits destined to interface at several different locations along the route. The SYNTRAN equipment can be provided with an optical or electrical interface so that it can be connected to a fiber optic cable or a twisted pair cable.

The second standard focuses on fiber optic networks above the DS-3 rate. This standard is SONET (synchronous optical network), which is based on a radically different network configuration requiring new equipment and restructuring of existing networks. Figure 17-23 illustrates how an optical interface based on SONET concepts fits into the network.

SONET defines a set of standards for a synchronous optical hierarchy that has the flexibility to transport many digital signals having different rates. This is achieved by defining a basic signal called the synchronous transport signal level 1

Figure 17-23   **Synchronous Optical Network (SONET)**

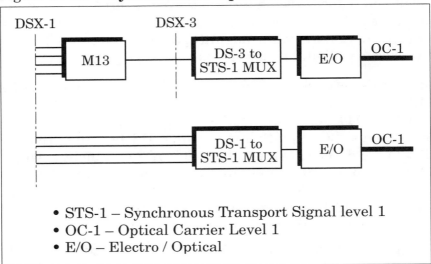

- STS-1 – Synchronous Transport Signal level 1
- OC-1 – Optical Carrier Level 1
- E/O – Electro / Optical

(STS-1) of 51.84 Mb/s. For transmission over fiber, an optical counterpart of the STS-1 signal, called the optical carrier level 1 (OC-1) is defined.

The OC-1 signal forms the basic SONET transmission building block from which higher-level signals are derived. Thus, an OC-3 signal operates at three times the OC-1 rate. The highest defined rate is OC-48, operating at 2488.32 Mb/s. In addition, work is underway on the next increase in the bit rate to OC-192, operating at 10 Gb/s. The STS-1 synchronous payload envelope shown in Figure 17-24 is nine rows by 87 columns consisting of 783 bytes. The byte from row one, column one, is transmitted first followed by row one, column two, and so on, from left to right, and from top to bottom. The transport overhead (27 bytes) contains the section overhead and line overhead, and it performs the function needed to transmit, monitor and manage the payload envelope over a fiber system. It also contains an STS-1 payload pointer, which indicates the start of the synchronous payload envelope. The payload envelope (783 bytes) consists of a path overhead (9 bytes) and a

Figure 17-24   **SONET STS-1 Transmission Format**

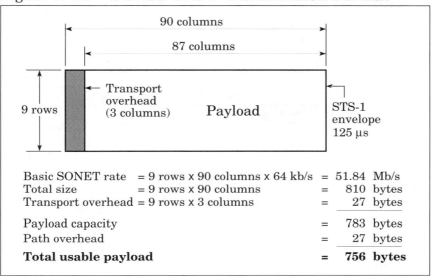

Basic SONET rate   = 9 rows x 90 columns x 64 kb/s = 51.84 Mb/s
Total size   = 9 rows x 90 columns   =   810 bytes
Transport overhead = 9 rows x 3 columns   =   27 bytes

Payload capacity   =   783 bytes
Path overhead   =   27 bytes

Total usable payload   = **756 bytes**

payload. The path overhead supports the monitoring and management of the payload. The payload carries the voice or data signals being transmitted. Several STS-1s can be combined for the transport of payloads in multiples of 51.84 Mb/s.

The STS-1 payload envelope can be divided into virtual tributaries (VTs) as shown in Figure 17-25. There are four sizes of VTs: VT 1.5, VT 2, VT 3 and VT 6, which have payload capacities of 1.5, 2, 3 and 6 Mb/s, respectively. Each VT has two parts, the VT payload pointer, which indicates the start of the VT payload envelope, and the VT synchronous payload envelope, which includes the VT path overhead and payload. As shown in Figure 17-26, higher transmission rates can be obtained by synchronously byte-interleaving "N" STS-1s to form an STS-N. N can be any of 1, 3, 9, 12, 18, 24, 36 or 48, as shown in Figure 17-27.

Since the 51.84 Mb/s rate does not handle the European hierarchy very well, the ITU-T (International Telecommunications Union-Telecommunication Standardization Sector) has agreed

Figure 17-25   **STS-1 Format**

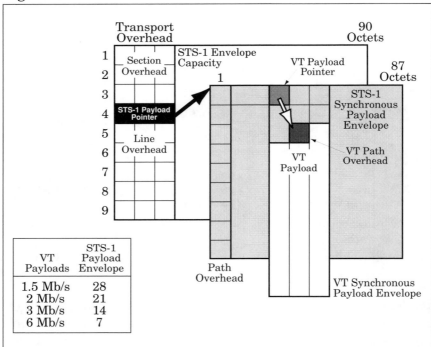

Figure 17-26   **SONET Multiplexing**

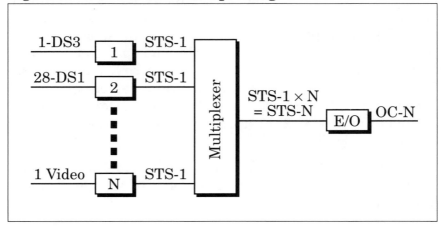

170

Figure 17-27   **SONET Optical Line Rates**

Optical Carrier Level	Electrical Equivalent	Line Rate Mb/s
OC-1	STS-1	51.84
OC-3	STS-3	155.52
OC-9	STS-9	466.56
OC-12	STS-12	622.08
OC-18	STS-18	933.12
OC-24	STS-24	1244.16
OC-36	STS-36	1866.24
OC-48	STS-48	2488.32

to an international standard based on a starting rate of 155.52 Mb/s, called the synchronous digital hierarchy (SDH). This will retain the OC-1 modularity, but below OC-3 the same interface will not be used worldwide.

A SONET point-to-point system is shown in Figure 17-28. The path overhead is carried within the payload envelope between the path terminating equipment across the SONET network. The line overhead operates between the line terminating equipment at each end of the line. The section overhead operates between the section terminating equipment, located at both ends of the section. A section is the fiber span between regenerators.

The implementation of the SONET and SDH standards has resulted in a complete family of lightwave, microwave and switching products which will create a global compatible network for the integrated delivery and management of communications services.

Figure 17-28   **SONET Point-to-Point System**

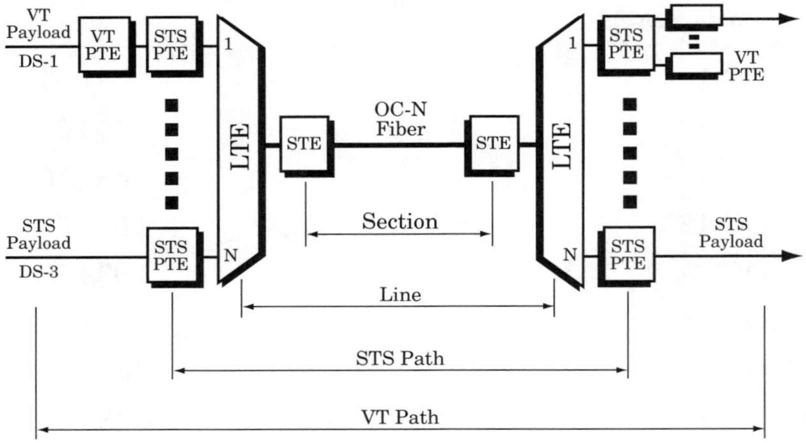

LTE = Line Terminating Equipment   PTE = Path Terminating Equipment
STE = Section Terminating Equipment   VT = Virtual Tributary

17.9   Protection Switching

With up to 32,000 voice circuits on each fiber pair, it is impera-
tive that a service outage does not occur on a long-haul lightwave system. In order to meet system availability objectives,
these systems require protection switching capability. When a
failure occurs, service is switched from the failed line to a
standby line. The switchover must occur automatically when
the bit error rate (BER) or some other parameter representa-
tive of a hard outage exceeds a particular threshold.

The switch is required to recognize a previous section failure
coming into the head-end of the switch and inhibit attempts to
activate switches beyond that point. A DS-3 pilot pattern or
"blue alarm signal" is normally required to be inserted on the
system to keep all downstream sections from alarming in the
event of a previous section failure. Remote control of the switch
is also provided as an option for high-capacity intertoll systems.

The system should be protected against equipment failures
on a 1:N basis, that is, one protection line for N regular working lines. The maximum value of N is determined by considering availability requirements, equipment reliability and
amount of traffic being carried. A typical arrangement is
shown in Figure 17-29.

Each direction of transmission must be separately protected
and separately powered at terminals and repeaters. Switching should also be available at the DS-3 level, with only the
degraded DS-3 being switched (see Figure 17-30). The switching operation is controlled by monitoring the performance of
the DS-3 signal and the line rate signal on the switching section. Initiation of switching based on BER performance at the
line rate is desirable due to the fact that the line rate is unique
to that section. However, if only the line error rate were used,
faults in the electronics between the line rate equipment and
the DS-3 rate equipment could go unnoticed. Thus, both the
line rate and DS-3 parity detection are required. The switch

Figure 17-29 **Protection Switching Arrangement at a Terminal Building**

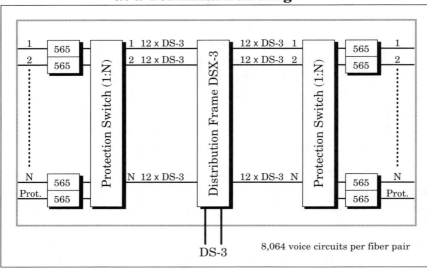

Figure 17-30 **Protection Switching 45 Mb/s (DS-3) Failure**

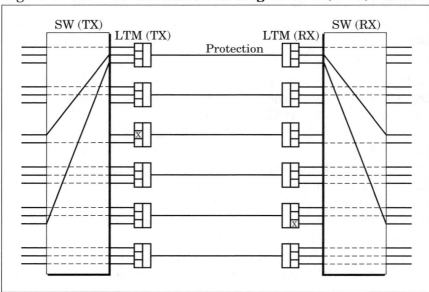

design should also allow a manual forced switch to take priority over automatic operation in the event of faulty automatic operation.

17.10   System Availability

All cable-based technologies are subject to cable outages from dig-ups, and lightwave systems are no exception. However, frequent catastrophic failures of a long-haul high capacity system could have a significant economic impact on the operating company and users alike. The time to restore a cable-related outage ranges between 6 and 12 hours depending on the nature and location of the break.

One of AT&T's major fiber cables between New York City and Newark, N.J., was cut in error by an AT&T workman on January 4, 1991. The outage lasted nearly 8 hours and wiped out 60% of the calls to and from New York City. Various New York stock and commodities exchanges were shut down for 3 hours. In addition, voice and radar data facilities at air-traffic control centers in New York City, Washington, D.C., and Boston were crippled for about 5 hours, grounding some aircraft.

Since the majority of catastrophic system failures have been caused by complete cable cuts (for example, construction dig-ups and vandalism), the selection of a secure right-of-way and the provision of some form of route diversity are the crucial elements in the survivability of a lightwave system.

Lightwave transmission systems, as with all transmission systems, are subject to two forms of outage sources, extrinsic and intrinsic. Extrinsic sources are divided into two categories: those which are avoidable (through careful system planning, design, installation and maintenance) and those which are unavoidable. Avoidable extrinsic failures are those failures that are due to improper engineering, installation or maintenance. Unavoidable extrinsic failures include natural disasters, cable dig-ups, damaged poles, and collapsing ducts. All types of cable, fiber or twisted pair, are subject to such failures. Intrinsic outage sources are those associated with the quality of the system hardware, including repeaters, terminals and cable, and are dependent on the manufacturer of that equipment. In general, the intrinsic reliability portion of lightwave systems is very good, primarily as the result of the use of automatic protection switching to guard against electronic component failures or single fiber failures. As a result, Section 17.11 will deal with cable placement techniques because the placement strategies are critical to system availability.

The placement method of fiber optic cables for long-haul systems will generally fall into one of three categories: underground ducts, aerial or direct buried. Each method of placement has some advantages when considering cost and system availability over various types of terrain.

The choice of which placement method is suitable is, to a large degree, dependent on terrain conditions. However, in areas where alternative placement methods are available, the following factors should be considered. In unwooded rural areas, aerial cable is not as likely to be subjected to intentional vandalism (such as a bullet or shotgun blast) due to the higher risk of individuals being identified and caught. Also, this same cable is less likely to be exposed to hazards of forest fires. Furthermore, in open rural areas, aerial cables are less likely than direct buried cables to be accidentally disturbed by farmers' tractors and backhoes, as a result of plowing, trenching and ditchdigging operations. Placement of aerial cable on existing pole lines has the obvious advantage of cost saving.

The major disadvantage of aerial cable is that of increased vulnerability to accidental and intentional cable damage as well as fire and storm damage in many areas.

As a result, route selection will heavily impact the operational availability of a lightwave system. For many areas there may be no "ultrasecure right-of-way." Of the various alternatives, the following are ranked in order of perceived desirability: restricted access highway, utility right-of-way, railway right-of-way, pipeline right-of-way and private easement.

Some of the lightwave systems have primarily used railroad rights-of-way as fiber optic corridors. Some of those networks are "single-thread" systems with no alternate routing in case of a catastrophic failure. Railway rights-of-way are subject to train derailments and construction dig-ups. An exploding chemical tank car could create a 25 foot deep crater.

The system design availability objectives of 99.985% for a 5,000 km system, comparable to a digital radio system, are realistic in terms of the terminal and repeater equipment. However, operating experience has indicated that the structural availability of the cable could be as low as 96.5%. Based on a 5,000 km system, the mean downtime (MDT) could be over 300 hours per year. An overall operational allocation for cable and equipment of 99.925% (7 hours/year) should be aimed for. The objectives for systems less than 5,000 km can be derived by prorating linearly.

Lightwave systems, unlike radio systems, do not present a finite upper limit to system capacity. However, the larger the system size, the more difficult it is to restore the system in the event of a catastrophic failure.

To overcome this situation, the system designer must either select a more secure route or configure the system to accommodate the expected outage through the use of diversity.

Smaller capacity systems operating at 140 Mb/s (three DS-3s) or less could use the protection channel on a parallel digital radio system for full or partial restoration. However, larger capacity systems operating at 565 Mb/s (12 DS-3s) or more would require a parallel SONET radio system or a dual cable path utilizing two cables with as much separation as possible between the cables.

Finally, emergency restoration procedures must be developed and maintained. Usually, this requires the provision of spare reels of cable and some mechanical connectors as a quick but temporary restoration measure.

17.11   Cable Placement Techniques

As noted in Section 17.10, the placement strategies for fiber optic cable are critical to the overall performance and availability of the transmission system.

Optical fiber cables are available for all conventional types of installations and environmental conditions required of twisted pair cables. These include in-building, underground conduit, aerial, direct buried and underwater installations.

The placement method selected will, of course, depend on the terrain involved and local circumstances. Ideally, all cable should be out of sight (that is, not aerial) to avoid intentional vandalism; however, high plowing costs or digging in many rocky or mountainous areas may make this impractical.

In general, no special placement precautions are required except to maintain the recommended pull tension and minimum cable bend radius. Because of the small size and low weight of optical cables, handling is easier than that of twisted pair or coaxial cables, and longer continuous lengths can be placed where there are no obstacles in the route. The long lengths reduce the number of splices in a route but require very careful planning of the cable placement process and routes.

• Placement of Underground Cable

Where it is desirable to place the cable below ground, the use of underground conduit is recommended. However, this placement method is generally more time consuming and expensive than direct burial. A typical underground duct-type cable is shown in Figure 17-31.

A pulling eye is readily attached to the strength member of the optical cable. A pulling rope or winch line is then connected to this eye with a swivel. The use of a sturdy pulling rope (at least one cm diameter), which will not stretch during the pull, is recommended. Although a steel wire rope would not stretch, it does have a tendency to cut into the duct if the rope diameter is small. A less sturdy rope will stretch during installation, causing surging problems at the cable reel.

During the pulling-in operation, the cable's pulling tension should not be allowed to exceed the value given in the cable specification. Typical underground cable ratings are in the 1,300 – 2,700 N (290 – 600 lb) range. If conventional cable placing equipment is used, tension monitoring is essential. Of the several methods available, the simplest is the use of a dynamometer.

Figure 17-31   **Underground Duct Type Cable**

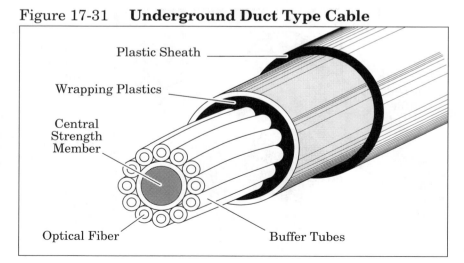

Plastic Sheath

Wrapping Plastics

Central Strength Member

Optical Fiber

Buffer Tubes

Because of long cable lengths, center pulling is frequently the preferred method of installation. The cable reel is placed at a middle access point and pulled in one direction. The remainder of the cable is then removed from the reel, laid out in a figure-8 pattern to preclude kinking, and pulled by the other end into the conduit in the opposite direction. The minimum bend radius should not exceed the value given in the cable specification when using pulling sheaves (typically 16x to 20x the cable outside diameter).

Since an underground system is the most expensive item in outside plant construction, every effort should be made to maximize the use of existing conduit facilities. One popular method involves placing three or four 1-inch plastic tubes (known as subduct, ductliner or inner-duct) in a standard 4-inch duct as shown in Figure 17-32. This provides a clean and isolated environment for each cable and permits additional cables to be pulled in at a later date without interfering with or damaging the existing cables.

Figure 17-32 **Subduct System for Fiber Optic Cables**

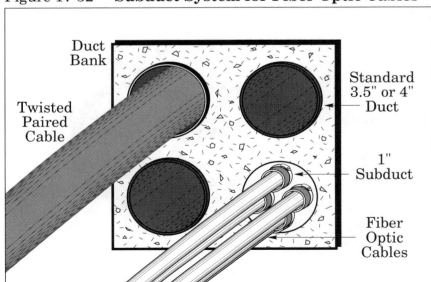

• Placement of Aerial Cable

The tension and bend radius limitations of fiber optic cables do not place a severe constraint on aerial placement. The preferred method is to lash the optical cable to a dedicated messenger wire or install a cable with its own messenger (figure-8 type) as shown in Figure 17-33.

The engineering rules for the aerial installation differ from those for conventional twisted pair cables in that fiber optic cables are designed for limited cable elongation.

In order to guarantee the long-term reliability of a fiber optic cable, one must know the amount of strain it experiences under various loading conditions (wind and ice), change in temperature and creep of the messenger wire.

Comparing the computed strain values to the maximum allowable strain in loose tube type cable designs, it has been concluded that these cables can be safely installed in the great majority of situations.

Aerial cables in continuous lengths up to 3 km have been placed by either the stationary-reel or moving-reel method of instal-

Figure 17-33   **Aerial Figure-8 Type Cable**

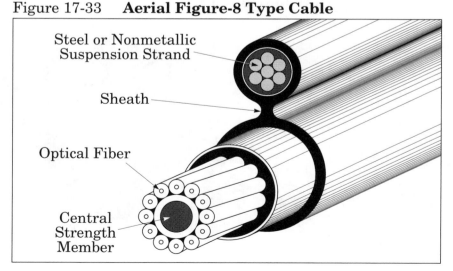

Steel or Nonmetallic Suspension Strand

Sheath

Optical Fiber

Central Strength Member

lation. It is recommended that a minimum clearance height of 5.4 meters (18 feet) be maintained to decrease the chance of accidental knockdown of the cable. In most public easements, the minimum height is dictated by the agency granting the easement. Also, the National Electrical Safety Code specifies minimum clearances in the United States.

Electric utilities have several options for installing fiber optic cables. They can use a self-supporting cable with a steel or nonmetallic strength messenger and install the cable on their poles or towers. Fiber cable can also be attached to an existing conductor. However, the most popular method is to install a new static wire that also contains the optical fibers.

Composite static wires are now available from a variety of cable manufacturers. These cables, called optical ground wires, are designed to replace the existing static wire on high voltage transmission lines. This cable, shown in Figure 17-34 can provide a cost-effective method of combining power transmission lines and telecommunication cables on the same secure right-of-way. This technique requires careful consideration of optical fiber maintenance methods and procedures.

Figure 17-34   Optical Ground Wire

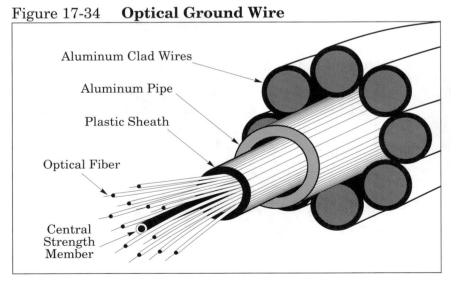

• Placement of Direct Buried Cable

Armored cables are preferred for direct buried applications. These cables can be placed by means of a conventional cable-laying plow of either the vibratory or static type. The cable reel is placed on the plow as is done with conventional cables. The optical cable is then guided over a deflector or a wheel and fed into the plow blade conduit.

All-dielectric unarmored fiber optic cables also have been plowed and trenched successfully. However, metal armoring, 0.125 mm thick or more, is necessary to provide protection against sharp rocks and rodents such as gophers. A typical armored cable is shown in Figure 17-35.

Figure 17-35   **Direct Buried Armored Cable**

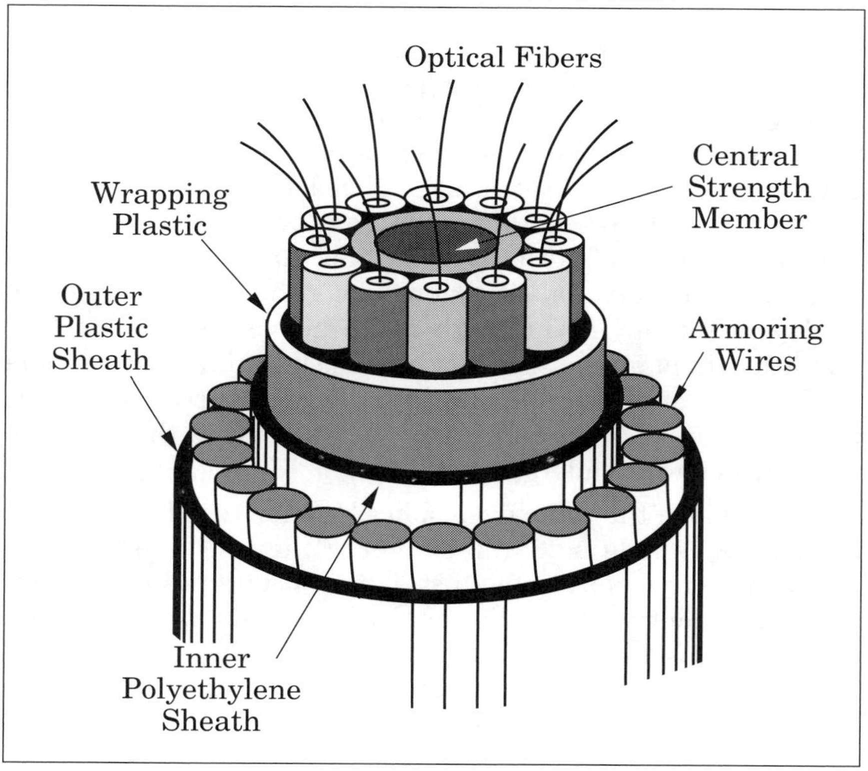

In practice, the reel lengths for direct buried lightwave projects are generally limited by terrain rather than by the maximum allowable pulling tension, as is the case when placing cable in conduit.

On buried routes, special construction methods are often required to negotiate natural or man-made obstacles, such as streams or roads. A rigid duct of suitable length and diameter should be placed beneath such obstructions beforehand. The cable is pulled through the duct using underground cable-placing techniques.

Generally, splices will be placed near obstacles which require a duct, because access to one end of the cable or the other is required to accomplish the pull into the duct. The number of splices necessary and the transmission design of the system, therefore, may depend on the nature of the terrain along the route.

A successful direct buried cable project depends on careful planning. A key element in planning is a careful route survey. The outside plant engineer and the construction supervisor who will be responsible for the cable installation should jointly carry out the route survey to establish splice locations, construction methods suitable for soil conditions, points of access to the right-of-way and depth of burial. During the survey, the two should note special or unusual situations along the route and plan to deal with them. A thorough route survey is particularly important in planning a project because it is almost impossible to abort a fiber optic cable plowing operation successfully once it has begun. Pre-ripping the route will ensure there are no unknown objects underground.

Each section of the route from splice location to splice location, a distance as long as 3.2 kilometers (2 miles), must be prepared properly before cable installation begins. It is very important to identify all conflicts and obstructions along the route as soon as possible. These situations may influence the selection of splice locations and thereby directly affect the overall transmission design of the system.

The cable should be buried to a minimum depth of 1.2 meters (4 ft) in long-haul applications. When roads or ditches are crossed, a concrete or steel pipe should be used. An optically visible marker tape, such as bright orange, should be placed directly above the cable approximately 0.15 to 0.3 meters (6 to 12 inches) below the surface. The warning tape or ribbon should be printed with a warning message and phone number for the cable owner. Also a similarly colored flag marker should be placed above the cable at the more vulnerable locations such as road crossings. The use of above-ground pedestals can be avoided by using below-ground splicing enclosures. These will normally be protected in some other type of enclosure or barrier. These splice locations must be locatable. This is often accomplished through the use of an "electronic locator peg" buried near the splice closure. These below-ground splices using filled cables can be equipped with splice moisture detectors capable of reporting to a centralized alarm system. All buried cables must be locatable. This is currently accomplished by detection of the metallic content of the cable. The metallic content can be in the sheath or the strength member. When all-dielectric cables are used, a #6 AWG insulated conductor may be buried along with the fiber cable for locating purposes.

One unique fiber optic corridor in the United States has been created by a pipeline company. A lightwave network has been built across the United States largely in empty pipelines. These deactivated 6 inch, 8 inch and 10 inch pipelines provide an extremely safe and secure environment for the cable. In addition, it is a highly economical method of installing cable because it avoids the traditional and costly method of digging and trenching. The cable was pulled through the pipes in lengths up to 3 miles long, which minimizes the number of required splices. Electric utilities and pipeline companies are now providing highly secure cable corridors which could increase the availability of long-haul lightwave systems.

As higher bit rates provide even greater system capacity, cable placement techniques and the selection of highly secure rights-of-way will become even more critical to lightwave system survivability.

17.12   Field Splicing

Two methods of permanently joining two fibers are being used successfully in the field: fusion splicing, using an electric arc to fuse the fibers, and mechanical splicing.

In the fusion technique, shown in Figure 17-36, the fiber ends to be joined are placed in physical contact and then heated past their melting temperature until they are fused together. The heat source is an electric arc controlled by an automated electric timer. The fusion procedure is sometimes conducted with

Figure 17-36   **Fusion Technique**

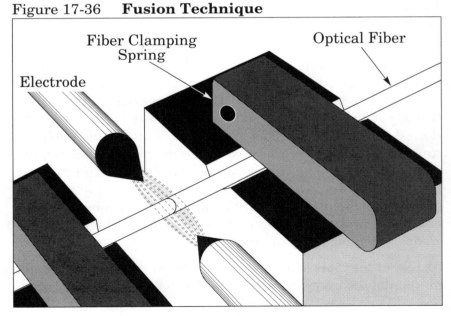

the aid of an integral microscope. However, many manufacturers use video projection techniques and LCD screens, which have proven to be very effective.

A splicing team typically consists of two craftsmen; a splicer and the test equipment operator, who monitors splice quality. The splicing sequence includes setup, opening cable sheaths, anchoring cables, fiber splicing (including testing), installation of closure, racking and packup.

The introduction of singlemode local injection-detection (LID) monitoring equipment has enabled both splicing time and splicing loss to be reduced. As shown in Figure 17-37, this consists of locally injecting a light source into the fiber and locally detecting on the other side of the splice point. A computer automatically adjusts the X and Y coordinates of the fibers, until maximum optical power transfer is achieved at the splice point. The splice is then fused with either an AC or DC arc. These methods can lead to average splice losses in the 0.03 to 0.07 dB range.

Although fiber optic splicing can be performed in manholes or on the aerial messenger, splicing in a splicing van has been a popular method. The resulting cable slack and splice closures are stored in a cable enclosure for protection.

Figure 17-37   **Local Injection & Detection**

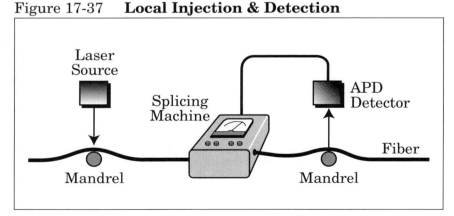

Most of the mechanical splices presently used can be grouped into three categories: epoxy and polish splices, cleaved splices using ultraviolet curable epoxies and mechanical cleave and crimp splices. A typical rotary splice uses the epoxy and polish method. They offer splice losses equal to that generally attainable with early fusion splicing. A typical elastomeric splice is ultraviolet curable with less termination time but at the expense of somewhat higher losses. A typical cleave and crimp splice requires much less sophistication on the part of the installer and eliminates the need for epoxies, but translates into a slightly higher loss. In practical systems, there is very little difference between mechanical splice types.

At the present time, fusion splicing is the most popular method in the telephone industry. Using well-trained technicians, high-quality splices have been obtained on a consistent basis. However, this could change in the future. Further advances in mechanical splice technology, lower costs and the provision of fiber in the local loop may provide a larger market for the mechanical splice.

17.13   Field Testing Using an OTDR

Field testing of fiber optic cable using an optical time domain reflectometer (OTDR) is usually performed in order to monitor the fiber splicing process to ensure that end-to-end transmission specifications are met and to provide baseline data for cable plant maintenance.   A typical OTDR is shown in Figure 17-38.

The OTDR has an important role to play not only during the installation and acceptance of fiber cable but also in the future maintenance of that cable.

The OTDR in essence is a one-dimensional closed-circuit optical radar. It operates by periodically launching narrow laser pulses into the fiber under test and monitoring the reflected light with respect to time. When light is launched into a fiber, reflected light is produced by two mechanisms: Fresnel reflection and Rayleigh scattering.

Fresnel reflection (see Figure 17-39) occurs as light passes into a medium having a different index of refraction (glass to air). A maximum reflection will occur if the fiber at the far end is well cleaved. Approximately 4% of the light within a fiber is reflected by an ideal cleave. In practice, however, fiber faults

Figure 17-38   **Typical OTDR**

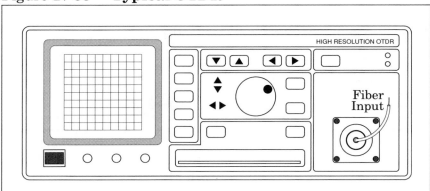

189

Figure 17-39 **Fresnel Reflection**

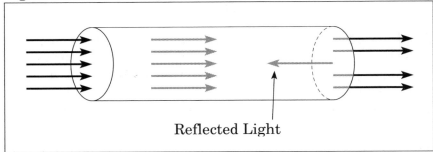

Reflected Light

or breaks are unlikely to produce an ideal cleave. Therefore, in these cases the Fresnel reflection will generally fall short of the maximum reflection.

Light traveling down a fiber is reflected throughout the length of the fiber by Rayleigh scattering. Rayleigh scattering (see Figure 17-40) is caused by light striking glass molecules and being scattered in all directions. The small amount of scattered light that goes back to the OTDR detector is called backscattered light. Since scattering is the most dominant loss mechanism in high-grade fibers, the backscattered light closely approximates the attenuation along the length of the fiber.

The OTDR uses the scattering properties of the optical fiber and allows the operator to "look into" the optical fiber. With

Figure 17-40 **Rayleigh Scattering**

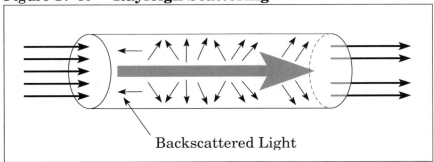

Backscattered Light

the OTDR, flaws and breaks can be found and localized, cable lengths determined, splice, connector and section losses identified and a picture of the attenuation versus distance down the cable obtained. In addition to the display on the display screen, some OTDRs have either a built-in chart recorder or provide plug-in access for a separate recorder. The recorder can measure in various units and marks at intervals. Figure 17-41 is a typical trace showing a splice as a vertical dip. The splice loss is the vertical distance (cm) multiplied by the chart recorder attenuation setting.

Markers are used to identify a position just to the right and left of the splice. The instrument then calculates the attenuation and displays the value on the display. Expansion controls in both the horizontal and vertical direction allow pinpointing of the points just prior and just after the splice.

Some OTDRs are designed in a modular package providing separate plug-in modules for 850, 1300 and 1550 nm wavelengths. There are many manufacturers producing OTDRs. The dynamic range of these instruments varies from 15 dB to

Figure 17-41   Chart Recording of a Fault or Splice

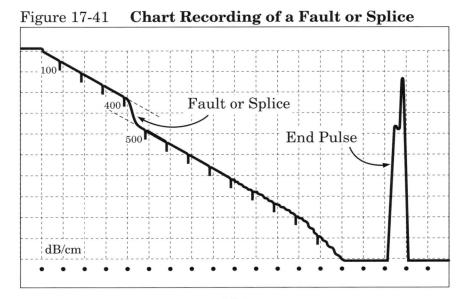

28 dB at 1300 and 1550 nm wavelengths. The dynamic range is defined as how far out you can see in a given fiber span.

The OTDR is used to locate any faults or breaks that may occur in a cable after it is in service. These faults will usually appear as a shift or reflections not present when the cable was originally installed.

The following is a summary of the tests performed by the OTDR on the cable.

- Prior to installation → Cable damage; Cable length on reel
- Splicing → → → → → Splice loss
- After installation → → End-to-end attenuation
- Maintenance → → → Fault locating

17.14   Fiber to the Home

The final frontier for lightwave systems is the so-called last mile of the network, the local loop between the central office and the customer's home or business. Telephone companies have begun to replace their aging twisted pair cables with fiber optic cables in the local loop. This will be a major break with "plain old telephone service" (POTS) that has characterized the telephone business for the past 100 years.

With this change will come the opportunity for the telephone companies to offer many new services that require the larger bandwidth. These services include broadcast cable TV and switched video. Instead of the 35 to 50 video channels brought into the home simultaneously on today's coaxial cable, a customer would select a channel from a large menu of possibilities, with switching done by the network. In fact, the complete replacement of coaxial systems with fiber optics could be driven by consumer demand for high definition television (HDTV). These new television receivers achieve higher quality pictures by increasing the scan lines from 525 to 1,125 per frame. The larger picture is so detailed and so fine-grained that it looks as sharp as a 35 mm color slide. These systems will require the larger bandwidth that only fiber can provide.

The near-term alternative is to provide fiber to the curb (FTTC) or distribution pedestal. This approach allows several customers to share the cost of expensive lasers, detectors and other electronics for the fiber system, while continuing to use their existing metallic drops.

Many fiber-based local loop systems now being tested use a star architecture. This is similar to the existing twisted pair network and require individual fibers to each customer from digital loop carrier terminals in each neighborhood. A less expensive alternative is a passive optical network architecture which is more reliable (fewer components) and easier to maintain.

As shown in Figure 17-42, in the existing twisted pair local loop system, PCM carrier is transmitted over twisted cable pairs between the remote terminal and the central office. The remote terminal demultiplexes the individual lines and converts them to analog. Multipair twisted pair cables distribute the analog signals from the remote terminals to the individual premises. Lines are dropped from the cable to the premises from service access points. To replace twisted pair cable with fiber, a fiber optic cable is used as the feeder from the central office to a host digital terminal (HDT) as shown in Figure 17-43. The HDT connects the distribution fiber network via the feeder network to the central office switch. Each HDT carries up to 384 voice circuits and can be installed in a remote cabinet hut, in a controlled environmental vault (CEV) or in the central office. The HDT operates over a passive optical network (PON) (see Figure 17-44) that splits the optical signal to serve up to 32 optical network units (ONU). The passive splitter is a 1 x N tree configuration, where N can be from 2 to 32. The ONU receives and transmits a packetized optical signal based on a specific ONU address in each packet. It performs the

Figure 17-42   **Twisted Pair Loop System**

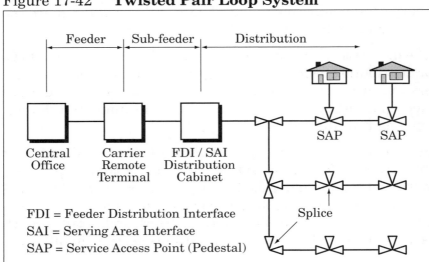

Figure 17-43   **Fiber Replaces Copper Cable**

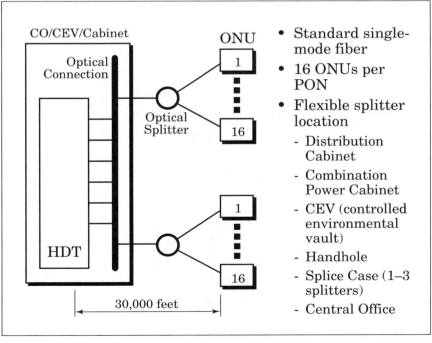

| Feeder | Sub-feeder | Distribution |

HDT

ONU   ONU

Central
Office

Host Digital Terminal
replaces
Carrier Remote Terminal
and automates
all distribution cabinets

Optical Network
Unit replaces
Service Access
Splice

Figure 17-44   **Passive Optical Network (PON)**

CO/CEV/Cabinet

Optical
Connection

Optical
Splitter

HDT

ONU
1

16

1

16

30,000 feet

- Standard single-mode fiber
- 16 ONUs per PON
- Flexible splitter location
  - Distribution Cabinet
  - Combination Power Cabinet
  - CEV (controlled environmental vault)
  - Handhole
  - Splice Case (1–3 splitters)
  - Central Office

electro-optical and digital to analog conversions and provides the interface between the loop fiber system and customers' metallic drops as shown in Figure 17-45. The ONU has mounting options for a pole, pedestal or handhole mount, and can serve up to 48 drops. However, most systems serve 4 to 8 drops.

The fiber-based network includes a telephone management software system which is used to maintain the optical network using remote monitoring and diagnostics. The network can be easily upgraded as new broadband services become available by utilizing wavelength division multiplexing along with an electronics upgrade.

Many of the fiber-in-the-loop (FITL) vendors claim that under certain deployment and volume production scenarios, FITL systems have achieved cost parity with twisted pairs. As a result, a number of telephone companies in the United States and Canada are beginning the massive job of replacing the twisted pair cables in the nation's local loops with fiber.

Figure 17-45   ONU Functions

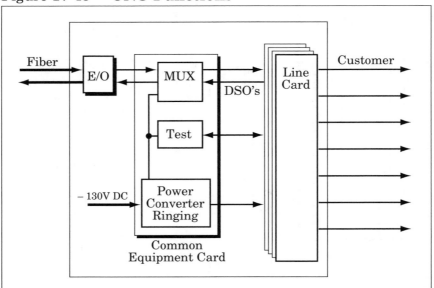

Conclusion

The principal elements of a lightwave system consist of a light source (the transmitter), a lightguide (the fiber), and a light detector (the receiver). The light source receives a coded electrical signal (containing voice, data or video), converts it to a light signal and launches this signal into the lightguide. The light signal is confined along the lightguide to the detector, where it is received and reconverted into the original electrical message.

The light source is either a light emitting diode (LED) or an injection laser diode (ILD). The light detector is either a positive-intrinsic-negative diode (PIN) or an avalanche photodiode (APD).

Lightwave systems are unaffected by environmental extremes and are immune to crosstalk and electromagnetic interference. Typical bit error rate performance is one error in 10^{11}. Therefore, the performance of lightwave systems is substantially superior to that of terrestrial microwave, satellite and twisted pair transmission systems.

All over the world, lightwave systems are being used in a wide variety of applications, including metropolitan telephone systems, long-haul trunking systems, undersea cable systems and local area networks and in the local loop to the home.

All-optical networks will widen the information superhighway of the future to cost effectively support the development of new high speed data services, and high bandwidth video and multimedia applications.

Review Questions for Section 17

1. Describe the three main parts of a fiber optic cable.
2. Name two types of optical fiber buffers.
3. Define numerical aperture (NA).
4. Describe the three transmission windows in a typical optical fiber.
5. Define multimode dispersion or pulse spreading.
6. What are the three types of fiber according to material composition?
7. Describe dispersion-shifted fiber.
8. Name five styles of fiber optic connectors.
9. Name two types of optical sources or emitters.
10. Name two types of optical detectors.
11. Briefly describe the difference between a FP and a DFB laser.
12. Identify the portion of the spectrum where most lightwave systems operate and why the light is not visible to the human eye.
13. How many voice circuits can a system carry on a pair of fibers operating at 565 Mb/s?
14. Define the bit error rate (BER) performance of a typical lightwave system.
15. Explain the reasons for developing a link loss budget.
16. Briefly describe the technique of wavelength division multiplexing (WDM).
17. Explain how an Erbium Doped Fiber amplifier works.
18. How many DS-0 (64 kb/s) bit streams are used to form a DS-1 signal?
19. At what rate does the optical carrier level 1 (OC-1)

operate within the SONET system?

20. What is the highest defined rate in the SONET hierarchy?

21. Define 1:N protection switching.

22. Explain briefly the subduct system for underground lightwave cables.

23. What is the minimum depth for direct buried lightwave cable in long-haul applications?

24. Briefly describe the fusion technique in field splicing of optical fibers.

25. Name the three categories of mechanical splices.

26. Briefly describe the operation of an OTDR.

27. Explain the difference between Fresnel reflection and Rayleigh scattering.

28. What is a PON?

29. Briefly describe the function of an optical network unit (ONU).

18 Local Area Network (LAN)

A local area network (LAN) is a communication network which is usually owned and operated by the business customer. A LAN enables many independent peripheral devices, such as terminals, to be linked to a network through which they can share expensive central processing units, memory banks and a variety of other resources.

These devices communicate over distances up to 2 km, for example, within an office building, a university campus, an industrial park or a hospital. Some organizations, such as banks, insurance and finance companies, have enough data traffic within a city to make intracity networking a viable means of reducing communication costs. In this case, individual LANs are interconnected to form a metropolitan area network (MAN).

Local area networks can be categorized according to their transmission bandwidth, either baseband or broadband. In baseband LANs, the entire bandwidth is used to transmit a single digital signal. In broadband networks, the capacity of the cable is divided into many channels which can transmit higher-speed signals. In the past, broadband networks have typically operated over coaxial cables.

A traditional approach for handling premises communications consists of a centrally located circuit switch, or private branch exchange (PBX), which connects pairs of intercommunicating devices for the duration of a call (refer to Section 12). While this approach is well suited to voice applications, it has several drawbacks for bursty data traffic. First, the capacity of the central switch is inefficiently used since most data transmissions consist of short bursts with long idle periods. Second, each port of the switch can only support a low fixed bit rate (typically 56 kb/s), thereby requiring that multiple ports be dedicated to devices demanding high-bit-rate service. Finally, circuit switching is not data-feature intensive; even simple functions, such as host multiplexing of several channels through a common host port, are not easily provided in a point-to-point circuit-switching environment.

A local area network overcomes the drawbacks by providing a high-speed channel that interconnects and is shared by all devices.

One of the most popular LANs in use today is Ethernet, shown in Figure 18-1. It was developed by Xerox and operates at 10 Mb/s over coaxial cable. Unshielded twisted pair wiring (UTP) is used on most new installations under the 10BaseT standard. Theoretically it allows an unlimited number of devices to be connected. But practical systems are limited to less than 100 – 200. A station with data to transmit first listens to the medium to determine if another transmission is in progress.

Figure 18-1   **Ethernet**

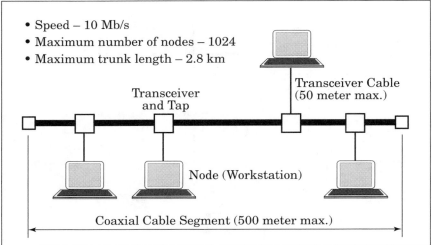

- Speed – 10 Mb/s
- Maximum number of nodes – 1024
- Maximum trunk length – 2.8 km

Transceiver Cable
(50 meter max.)

Transceiver
and Tap

Node (Workstation)

Coaxial Cable Segment (500 meter max.)

When the medium is clear, a station may start transmitting. If two stations attempt nearly simultaneous transmission, they will detect a garble or collision. The affected stations then cut short their transmission and wait a period of time before attempting retransmission. This access method is called carrier-sensing multiple access with collision detection (CSMA / CD).

IBM's token ring uses shielded twisted pair wiring or coaxial cable operating at 4 Mb/s and 16 Mb/s (see Figure 18-2). However, UTP may be used for both token ring LANs. When a station has data to transmit, it listens to the ring until it detects a pattern denoting the token. The station removes the token from the ring and transmits a frame of data. When the frame has circulated and returned, the station places a token back on the ring. The token will circulate until it encounters a station with data to transmit.

This technology is limited to 260 devices per ring with 12 rings, for a maximum upper limit of approximately 3,000 devices.

Over the past few years, interest in fiber optic local area networks has increased dramatically. Although the advantages of

Figure 18-2   **IBM Token Ring**

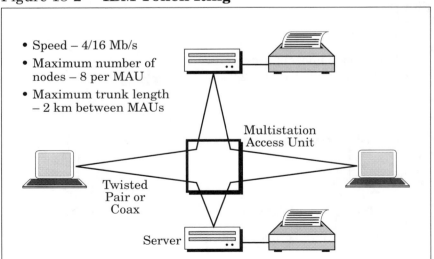

- Speed – 4/16 Mb/s
- Maximum number of nodes – 8 per MAU
- Maximum trunk length – 2 km between MAUs

Multistation Access Unit

Twisted Pair or Coax

Server

fiber optics are well known, it has been too expensive to use in LANs until very recently. However, the ever-increasing demand for optoelectronic components and the large-scale introduction of optical fibers in interoffice and long-haul portions of the conventional telecommunications networks have reduced the cost of fiber optic hardware to the point where it is now an economical alternative, especially in building backbone applications.

The fiber optic systems available for immediate LAN applications are based on a mature optoelectronic technology that operates at optical wavelengths of either 850 or 1300 nm. This technology has generated a family of reliable, inexpensive components that can be applied economically to LANs. State-of-the-art components, such as lasers and photodiodes, which operate at 1550 nm, are used in large-capacity long-haul systems but are not economical for use in most LANs.

62.5 micron multimode fiber is endorsed as a standard for LANs whereas 8 – 10 micron singlemode is used in the local loop and long haul systems.

Physically, LAN topology may take one of several forms. The most common ones are star, bus and ring. A star consists of long point-to-point links that carry traffic to and from a central node (see Figure 18-3). A physical star topology has been standardized.

In a passive star configuration, optical signals from all stations are guided to the central node, and a passive power splitter sends the signals back to all stations simultaneously in a multipoint or broadcast fashion. Star topologies can, in principle, accommodate hundreds of access stations.

The bus configuration, commonly used for LANs such as Ethernet, employs a linear length of cable (see Figure 18-4). Each station is connected to the cable by a tap. When coaxial cable is used, this connection is made by a high-impedance passive tap, and signals travel in both directions from access points.

Figure 18-3   **Star**

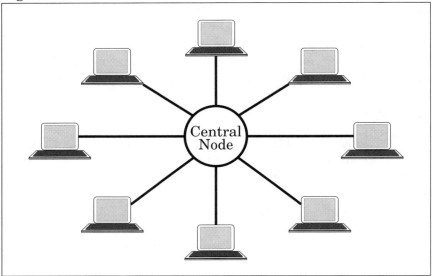

Figure 18-4   **Bus**

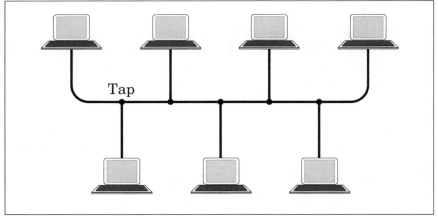

When optical fiber is employed, each station is connected to the cable by a bidirectional optical tap. The tap takes in a fraction of the power on the bus and inserts power bidirectionally. This ensures that the signal will reach all stations.

This power splitting, or optical tapping, limits the number of nodes that a fiber LAN can accommodate.

One solution to this problem is to close the bus back into itself to form a ring or loop (see Figure 18-5). In this way, communication can occur in one direction only. The nodes can be made into fully regenerative repeaters, called active nodes, that are connected by a simple point-to-point link. This eliminates the need for additional optical taps and splitters. A bypass mechanism is required for added reliability, and to allow devices to be disconnected for servicing.

Most of the local area networks installed to date have moderate data transmission speeds of 1 Mb/s to 15 Mb/s (roughly 400 to 7,000 typewritten pages per second), but in the near future LAN systems will require data rates of 200 Mb/s or more.

One manufacturer has recently announced a new wireless LAN product based on standard 18 GHz microwave technology. The new system is designed to extend or replace existing hard-wired networks such as Ethernet or token ring, as shown in

Figure 18-5   **Ring**

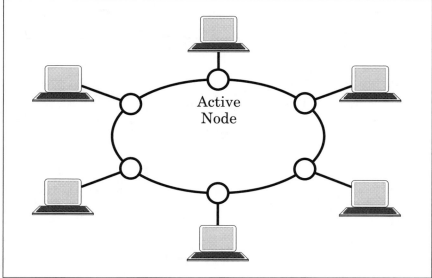

Figure 18-6. The system will be capable of initially delivering 15 Mb/s to the desktop and replacing the last 30 meters of wiring typically found in today's LAN installations. A ceiling mounted transceiver interconnects up to 32 computers and printers in a cell about 25 meters in diameter. The system eliminates the cost of wiring an office and, more importantly, the cost of changing the wiring when an office is reconfigured or moved.

Some fiber-based LANs use a time division multiplexing (TDM) ring (see Figure 18-7). Each node forms a drop and insert point for 1.5 Mb/s signals originating on the local systems. The TDM ring is based on a redundant counterrotating fiber ring of 45 Mb/s. It provides transport for either voice or data in standard digital 1.5 Mb/s aggregates. It can survive a cable cut anywhere along its path and still provide transport to all nodes.

Figure 18-6   **Wireless In-Building Network**

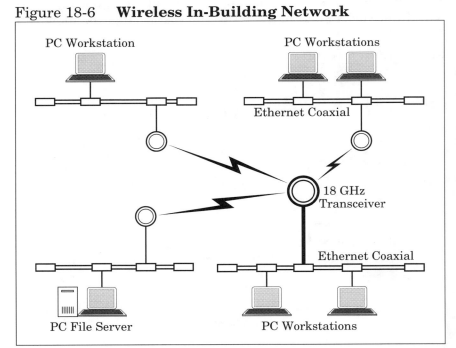

Figure 18-7   **Four Node TDM Ring**

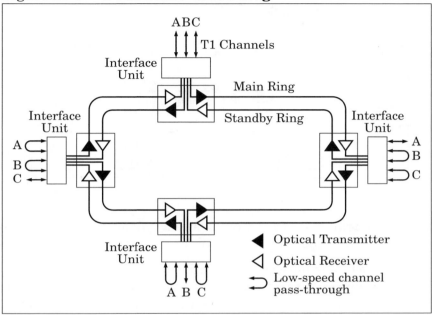

Each node uses an M13 multiplexer to bring the 1.5 Mb/s signals up to the 45 Mb/s rate and provide the electrical / optical interface.

For users with a need for very high-speed data transmission, a standard has been developed called a fiber distributed data interface (FDDI). FDDI is a frame-based technology that uses variable length frames to transport information across the network with a data rate of 100 Mb/s. As shown in Figure 18-8, the FDDI network is based on a backbone of dual, counter-rotating fiber optic rings. However, a standard has been developed to provide FDDI over unshielded twisted pair wiring in order to lower the cost for a workstation interface.

Figure 18-9 shows a total fiber distribution system that could be installed in a new building with terminals connected to wall outlets containing the electro-optical interface, which, in turn, would be connected to interconnect centers mounted in wiring

Figure 18-8   **Fiber Distributed Data Interface Network (FDDI)**

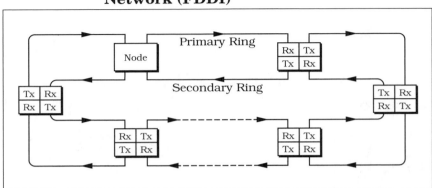

closets. The distribution panels on each floor would be interconnected by a fiber optic riser cable.

The riser cable is designed to go all the way from the building entrance splice to the user's desktop equipment in both the horizontal distribution and vertical riser. The riser cable, shown in Figure 18-10, is available with 62.5 / 125 micron multimode fiber or singlemode 8 / 125 micron fiber or both. Cable counts of 4 to 96 are available. The coated fiber is buffered to 900 micron with polyvinyl chloride (PVC). The PVC buffer is color coded or numbered for identification.

Interconnect or jumper cables used to link optical equipment are available in simplex or duplex styles (see Figure 18-11). Both use multimode 62.5 / 125 micron fiber operating at 850 nm or 1300 nm with 900 micron buffering of PVC. The cables are also available with singlemode 8 / 125 micron fiber operating at 1300 and 1550 nm buffered to 900 micron with PVC. The buffered fibers are surrounded by aramid yarn for strength and over jacketed with PVC for protection.

Fiber optic technology for LAN applications is maturing rapidly and there are a number of commercial fiber optic LANs and building distribution systems available to make "Fiber to the Desk" a reality. The full potential of optical fibers for local

Figure 18-9 **Fiber Distribution System**

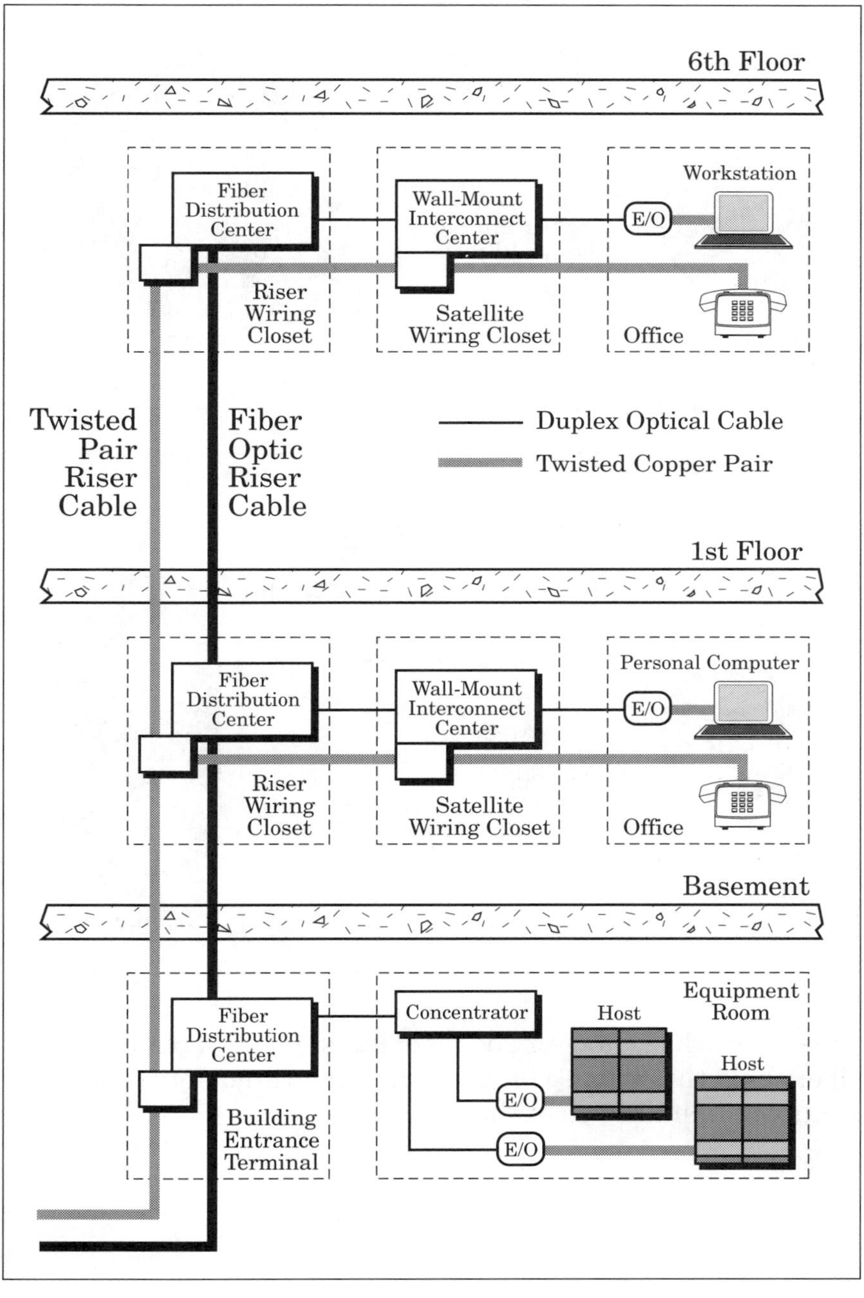

Figure 18-10   **Riser Cable**

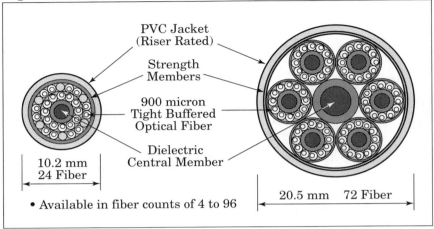

PVC Jacket
(Riser Rated)

Strength
Members

900 micron
Tight Buffered
Optical Fiber

Dielectric
Central Member

10.2 mm
24 Fiber

• Available in fiber counts of 4 to 96

20.5 mm   72 Fiber

Figure 18-11   **Interconnect Cables**

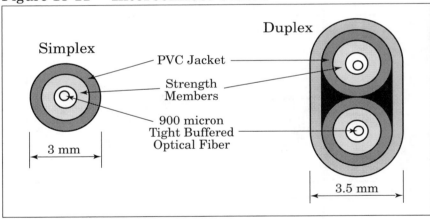

Duplex

Simplex

PVC Jacket

Strength
Members

900 micron
Tight Buffered
Optical Fiber

3 mm

3.5 mm

area networks is far from being exhausted. As a result, optical
fibers hold the promise for powerful integrated networks in the
commercial buildings of the future.

18.1   Integrated Services Digital Network (ISDN)

The next major step in digitization will be the worldwide integrated services digital network (ISDN). ISDN will provide direct digital services to the business or residential customer through a basic 144 kb/s interface with three channels. Two B channels will carry 64 kb/s transmissions of voice, high-speed data, graphics, facsimile and highly compressed video. The third, a 16 kb/s D channel, will carry control information and data.

The four principal categories of ISDN interfaces, as shown in Figure 18-12, are the basic interface structure (2B + D), the primary rate B (23B + D), the primary rate H0 (3H0 + D) and the primary rate H1 (nonchannelized).

Figure 18-12   **ISDN Basic and Primary Interfaces**

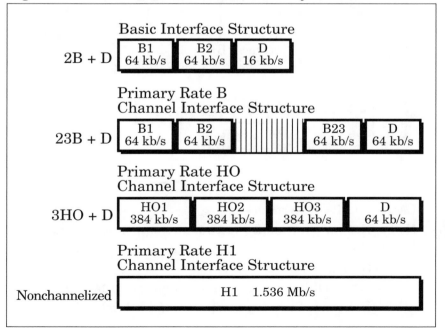

The potential ISDN digital service offerings will range from very low-rate telemetry (meter readings, security, energy management, etc.) to high-speed data and compressed video.

The basic requirements for ISDN user access have been defined by anticipating likely configurations of user equipment and interfaces. A typical user access model with definitions of terminations and reference points is shown in Figure 18-13.

The customer's personal computer would have the capability to permit two people connected through the ISDN, working at home or the office, to have a face-to-face conversation, review

Figure 18-13   Typical ISDN User Access Model

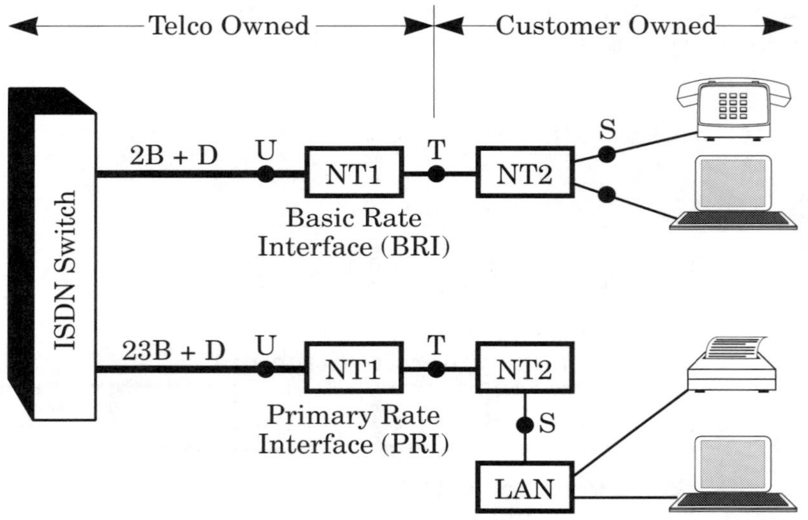

LAN – Local Area Network
NT1 – Network Terminal 1 (terminates transmission line)
NT2 – Network Terminal 2 (connection for multiple terminals)
 S – Switched Reference Point
 T – Terminal Reference Point (customer interface)
 U – Reference Point (transmission line to local switch)

a document, graphics or photos, then transmit a facsimile page in 5 seconds. The technology described in Section 15 is an example of the multimedia computers which would be used for this purpose.

ISDN is the result of more than fifteen years of effort to develop standards and concepts for a worldwide digital system that will provide networks that control a very broad range of voice and nonvoice services. The system will provide the flexibility to handle new services as they are developed. ISDN has been introduced in a number of countries, but growth to date has been slow.

Initially, one twisted pair of copper wires is adequate for all these services between the central office and the user's premises. Two cable pairs are required on the user's premises between the NT1 and the user's terminal.

Conclusion

A local area network (LAN) is a system which permits personal computers, workstations and other peripheral devices to share expensive resources, such as high-speed printers and large disk memories. Using various topologies such as the star, bus and ring, all stations are connected together by twisted pairs, coaxial cable or optical fiber. Individual LANs can also be interconnected within a city to form a metropolition area network (MAN).

The integrated services digital network (ISDN) will be a worldwide digital network for voice, data and compressed video. It will not be a separate new network, but will emerge from the gradual enhancement of existing networks and services. ISDN will provide digital transmission and switching directly to a standard outlet on the customer's premises over a twisted pair. The narrowband ISDN will be part of a graceful evolution toward a broadband integrated services digital network (BISDN) that will support all services.

Review Questions for Section 18

1. What is the approximate maximum length (km) of most local area networks?

2. What is the speed (Mb/s) of Ethernet?

3. Which transmission medium is normally used for Ethernet networks?

4. Does IBM's token ring network operate over twisted pair wiring or coaxial cable?

5. Which topology does Ethernet use?

6. What is the main advantage of a bidirectional ring over a unidirectional ring topology?

7. What does FDDI stand for?

8. What is the speed (Mb/s) of FDDI?

9. What is a riser cable?

10. What are fiber interconnect cables used for?

11. What is the size and material of the buffer used on 62.5 / 125 and 8 / 125 micron fiber?

12. Explain the basic interface structure of 2B+D.

13. What is the maximum speed (Mb/s) of ISDN operating over one twisted pair?

19 Broadband Integrated Services Digital Network

The worldwide telephone network is evolving toward a public system of broadband communications, integrating telephone, video and data communications. The future broadband integrated services digital network (BISDN) will be determined by the services that it provides. Some of the elements that are necessary to construct this network are now becoming available. A major element is the provision of high-capacity fiber optic cable in the trunk network and the local loop. Broadband is defined as the provision of customer access at bit rates in excess of 1.5 Mb/s. Broadband services are divided into two major groups, interactive and distributive services, as shown in Table 19-1.

Table 19-1

Interactive Conversational (point-to-point or multicast)	• video telephone • video conferencing • video / audio transmission • high-speed data transmission (LAN & MAN) • data file transfer • high-speed telefax • high-speed image transmission
Interactive Messaging and Retrieval	• video & sound • video mail • document mail • graphics, still images, motion video
Distributive	• broadcast TV • high definition TV • hi-fi sound

Due to the wide range of services that BISDN must support, a new transmission and switching technique called asynchronous transfer mode (ATM) has been developed. ATM combines the capabilities of time division multiplexing (TDM) and packet switching. It is a method of transferring digital information in the form of cells. Cells are of constant length regardless of the information carried. The cell consists of two parts, the header or label and the information field or payload.

Each cell consists of 53 bytes (octets) and contains a header (5 octets) and the information field (48 octets) as shown in Figure 19-1. Data carried in the information field could be from any source and operate at any speed. Pure data sources can produce very long messages, up to 64 kilobytes in many cases. By forcing these messages to be segmented into short cells, ATM ensures that voice and video traffic can be given priority and never have to wait more than one 53 octet cell time (3 microseconds at a 155 Mb/s rate), before it can gain access to the

Figure 19-1   **ATM Cell**

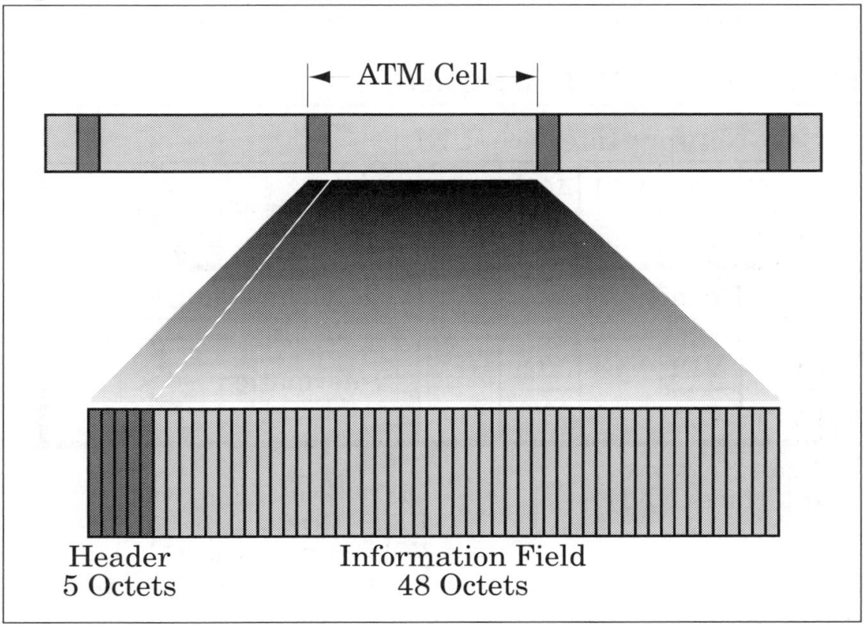

Header   Information Field
5 Octets   48 Octets

channel. With frame relay, in contrast, the wait would be a random interval of up to several milliseconds in length. The header contains an 8-bit error detecting code which provides protection against errors introduced by the network. With the extensive use of singlemode fiber in the network, errors will be so rare that more complex procedures are pointless. The header also contains a label field, which associates the cell with the service using that cell. The label field, in turn, contains a routing field, which is divided in two parts as shown in Figure 19-2:

- the virtual channel identifier (VCI) which associates each particular cell with a virtual channel

- the virtual path identifier (VPI) which allows groups of virtual channels to be handled as a single entity

The network node interface (NNI) is the interface between the transmission facility and the network node. The user network interface (UNI) is the interface between the user equipment and the network termination. There are two variants of the ATM cell structure, one for the UNI and one for the NNI. The generic flow control (GFC) field exists only at the UNI, where it is concerned with flow at the user premises end of the access

Figure 19-2   ATM Cell Structure

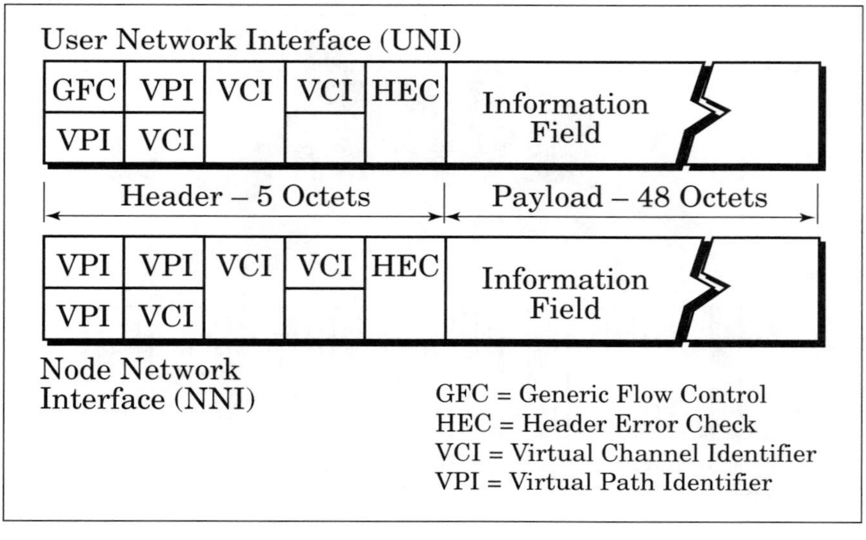

User Network Interface (UNI)

| GFC | VPI | VCI | VCI | HEC | Information |
| VPI | VCI | | | | Field |

Header – 5 Octets | Payload – 48 Octets

| VPI | VPI | VCI | VCI | HEC | Information |
| VPI | VCI | | | | Field |

Node Network Interface (NNI)

GFC = Generic Flow Control
HEC = Header Error Check
VCI = Virtual Channel Identifier
VPI = Virtual Path Identifier

link. The four bits used for GFC at the UNI are reassigned for the NNI header to extend the virtual path identifier at the NNI. Virtual channels (VC) are bundled into virtual paths (VP), as shown in Figure 19-3. Virtual channels VC1 and VC2 are bundled into VP1 on the link between node A and node B. Between node B and C, VC1 and VC2 are bundled into VP2. VC3 and VC4 are bundled into VP3 and carried directly over a different link to node C. A virtual path may consist of many virtual channels. Therefore the VP1 might represent a trunk between two cities and the VC1s might represent individual calls. Switching equipment along the way can route all calls on the basis of just the first octet of the address (VP1) without having to bother with the rest of the address until the trunk gets to the final location, where the traffic is distributed.

As shown in Figure 19-4, ATM cells can be mapped into SONET for scalable bandwidth channels with capacities into the Gb/s levels to support the huge amounts of data required for imaging, multimedia, interactive video and entertainment video applications. Over SONET, an ATM signal can be carried and multiplexed alongside payloads that use synchronous services such

Figure 19-3   Virtual Paths and Virtual Channels

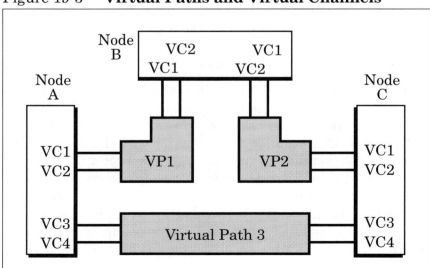

Figure 19-4   **ATM / SONET Mapping**

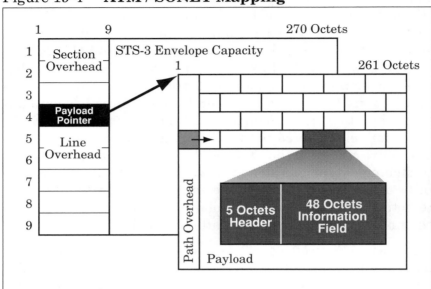

as DS-1 or DS-3. The network can carry these services on the same fiber without converting them to an ATM format.

ATM has become the basis for the transfer of information within all broadband networks. It is the technology used in switched multimegabit data service (SMDS), a public packet switched service now being offered by local exchange and interexchange carriers and will be the predecessor to BISDN.

SMDS provides connectionless (having no need of fixed or virtual paths) packet transport in fixed-length cells within variable length data units. Each data unit encapsulating up to 9,188 octets of information as shown in Figure 19-5. The 9,188 octet information field is large enough to accept a full frame of any existing local area network. Since it is compatible with international wideband standards, SMDS will run over the same ATM switches that will provide BISDN. The access speeds are presently at DS-1 and DS-3 but with an increase to a SONET rate of 155 Mb/s promised for the future.

Figure 19-5   **SMDS Cell**

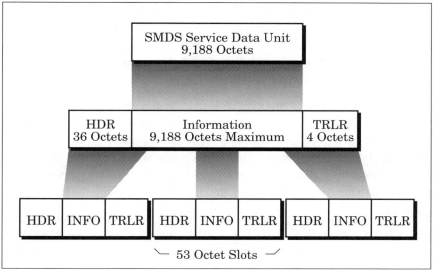

ATM is the first technology that promises to merge local area networks (LANs), metropolitan area networks (MANs) and wide area networks (WANs) into a single, simplified network. Most networks usually tie rates and formats firmly together. Ethernet operates at a rate of 10Mb/s, DS-1 at 1.544 Mb/s and so forth. The ATM standard covers only the 53 octet cell format, without specifying rates. So ATM cells can operate at common LAN speeds like 100 Mb/s or public network speeds of 155 Mb/s or higher. As a result, ATM users can send as many or as few cells as necessary to transfer their data. In addition, they pay only for the cells they send, not for the speed of a dedicated facility that may be used only part of the time.

A broadband network will include a number of nodes, cross-connect points and switches where cells are switched. A switch making use of the full combination of VP1 and VC1 is called an ATM switch, while a switch that uses only the VP1 is called an ATM cross connect. All cells with the same address take the same path from end to end. Unassigned cells are eliminated at the input and do not load the switching matrices.

ATM connections exist only as sets of routing tables in each switch, based on the address contained in the cell header. This information is used to direct a cell through the switch. The switch is therefore able to distinguish payloads that cannot tolerate delays (such as voice) from payloads that are bursty and less time-sensitive (such as data).

ATM concentrates bursty data (such as file transfer) from multiple sources and multiplexes these cells with synchronous information (such as video and voice) that has stringent end-to-end limits on delay variation. As shown in Figure 19-6, the transmission of delay-sensitive video cells is unaltered while some of the bursty data cells are delayed.

SONET transmission networks will provide high-speed optical transmission for voice, data, image and video between ATM switches. The information can be transmitted across SONET networks at rates of 155 Mb/s (OC-3), 622 Mb/s (OC-12) or 2.4 Gb/s (OC-48).

At some point, the customer premises equipment must intersect with the ATM network. This juncture is referred to as

Figure 19-6   **ATM Switch**

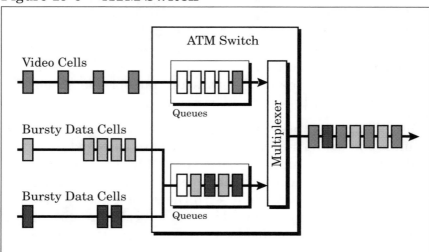

network termination 1 (NT1) and is located on the customer's premises. The interface will provide customer access at speeds up to 155 Mb/s and 622 Mb/s. Figure 19-7 shows one possible evolutionary scenario.

Telecommunications, multimedia computers, desktop publishing and entertainment systems are blending together to create sophisticated home and office communication. Lightwave, software and video technologies are making all of this possible.

Figure 19-7   **Broadband Integrated Services Digital Network**

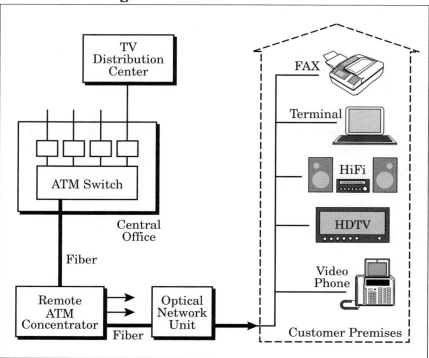

223

Conclusion

Telephone network planners have long dreamed of a single, ubiquitous, high-speed network capable of transporting voice, data and video directly to the customer. ATM, with its flexible cell-by-cell multiplexing, appears to be the best technology to achieve this dream. The main routes of this information superhighway will certainly make use of this technology, but the use of ATM cells all the way to the desktop will also be possible. This will result in a common environment that will encourage vendors to create applications that link users both locally and globally.

Multimedia, desktop videoconferencing, video-on-demand, all require the speed of ATM and its ability to handle a mixture of traffic. ATM not only provides a means of integrating voice, data and video, but it also joins local and wide area networks into a seamless whole.

The intelligent broadband information superhighway that was only a dream a few years ago is now a certain part of our future. The introduction of these new technologies throughout the public switched network will revolutionize communications and change the economic and social fabric of the whole world.

Telecommunications is the driving force that is creating a huge global economy. It will provide the infrastructure that every industry, company and individual will need to compete in a truly cosmopolitan global market place.

Review Questions for Section 19

1. Name the two major groups that make up broadband services.

2. How many bytes or octets are there in an ATM cell?

3. What is a VCI and a VPI?

4. Define the network node interface.

5. Define the user network interface.

6. What does SMDS stand for?

7. What is the maximum number of octets in the data unit of SMDS?

8. What are the access speeds for SMDS?

9. Briefly describe the difference between an ATM switch and an ATM cross connect.

10. Which ATM payloads are more delay sensitive, video or data?

11. At what optical rates will SONET provide transport for voice, data and video between ATM switches?

Telecommunications Glossary

Amplifier
A device for increasing the power of a signal.

Amplitude
The amount of variation of an alternating waveform from its zero value.

Amplitude Modulation (AM)
The process by which a continuous waveform is caused to vary in amplitude by the action of another wave containing information.

Analog
In communications, the description of the continuous wave or signal (such as the human voice) for which conventional telephone lines are designed.

Asynchronous Transmission
A transmission method in which each character of information is individually synchronized, usually by the use of "start" and "stop" elements (compare with "synchronous transmission").

Attenuation
A decrease in the power of a signal while being transmitted between points. Usually measured in decibels.

Bandwidth
The difference between the high and low frequencies of a transmission band, expressed in Hertz.

Baseband
In analog terms, the original bandwidth of a signal from a device (i.e., 4 kHz for telephone, 4.5 MHz for television).

Baud

A unit of signaling speed. Speed as expressed in bauds is equal to the number of signaling elements per second. At low speeds (under 300 b/s), bits per second and baud are the same. But as speed increases, baud is different from bits per second because several bits are typically encoded per signal element.

Bit

A contraction of binary digit, the smallest unit of information in a binary system of notation. Data bits are used in combination to form characters; framing bits are used for parity, transmission synchronization and so on.

Broadband (Wideband) Channel

A communication channel with a bandwidth larger than that required for baseband transmission. Very often any channel wider than voice grade is considered to be a broadband channel.

b/s

Bits per second (also expressed as bps) is a measure of speed in serial transmission. Also used to describe hardware capabilities, as in a 9,600 b/s modem.

Byte

A group of 8 bits. Often used to represent a character. Also called an octet.

CD ROM

Compact disc read only memory, a laser encoded disc that stores 650 megabytes of randomly accessible text, imagery and/or sound data.

Channel

In communications, a path for transmission (usually one way) between two or more points. Through multiplexing, several channels may share common equipment.

Character

Any coded representation of an alphabet letter, numerical digit or special symbol.

Clock

In data communications, a device that generates precisely spaced timing pulses (or the pulses themselves) used for synchronizing transmissions and recording elapsed times.

Coaxial Cable

A cable which consists of an outer conductor concentric with an inner conductor; the two are separated from each other by insulating material.

Code

A specific way of using symbols and rules to represent information.

Codec

An electronic device which converts analog signals to digital form, and back again. **Codec** stands for **co**der / **dec**oder.

Codec, Video

A codec which converts an analog video signal to digital form, and back again. In addition, video codecs generally compress or reduce the data rate to that which can be carried on a narrowband channel.

CPU

Central Processing Unit – the part of a computer that includes the circuitry for interpreting and executing instructions.

Crosstalk

Interference or an unwanted signal from one transmission circuit, detected on another (usually parallel) circuit.

CRT

Cathode Ray Tube — an electronic vacuum tube, such as a television picture tube, that can be used to display images.

Data

A representation of facts, concepts or instructions in a formalized manner suitable for communication, interpretation or processing; any representations, such as characters, to which meaning may be assigned.

Data Collection

The act of bringing data from one or more points to a central point.

Data Communications

The movement of encoded information by an electrical transmission system. The transmission of data from one point to another.

Dataphone

A trademark of the AT&T Company to identify the data sets manufactured and supplied by it.

Data Processing (synonymous with Information Processing)

The execution of a systematic sequence of operations performed upon data.

Data Processing System

A network of machine components capable of accepting information, processing it according to a plan and producing the desired results.

Data Set

A device containing the electrical circuitry necessary to connect data processing equipment to a communications channel, usually through modulation and demodulation of the signal.

Data Sink

The equipment which accepts the transmitted data.

Data Source

The equipment which supplies the data signals to be transmitted.

Data Stream

Generally, the flow of information being transmitted in a communications system or path.

dBm

A measure of power in communications: the decibel referenced to 1 milliwatt: 0 dBm = 1 milliwatt, with a logarithmic relationship as the values increase or decrease.

DDS
Dataphone® Digital Service, or dedicated digital service, a private-line service for digital data communications.

Decibel (dB)
A unit for stating the logarithmic ratio between two amounts of power. It can be used to express gain or loss without reference to absolute quantities.

Dedicated Link
A leased telephone line, reserved for the exclusive use of one customer; private line; special access.

Demodulation
The process of retrieving an original signal from a modulated carrier wave. The technique used in data sets to make communication signals compatible with business machine signals.

Detector
The means (usually PIN or APD) used to convert an optical signal to an electrical signal.

Diagnostics
The detection and isolation of a malfunction or mistake in a communications device, network or system.

Digital
In data communications, the description of the binary (off/on) output of computer or terminal. Modems convert the digital signals into analog waves for transmission over conventional analog telephone lines.

Distortion
The unwanted change in waveform that occurs between two points in a transmission system.

Distributed Processing
A general term usually referring to the use of intelligent or programmable terminals for processing at sites remote from a company's main computer facility.

EIA Interface

A standardized set of signal characteristics (time duration, voltage and current) specified by the Electronic Industries Association.

Emitter

The means (usually LED or laser) used to convert an electrical signal into an optical signal for transmission by an optical waveguide.

Erbium Doped Optical Fiber Amplifier

An optical amplifier utilizing a section of optical fiber doped with the rare earth erbium and optically pumped with a laser diode. It can amplify a range of wavelengths at the same time.

Error Rate

The ratio of incorrectly received data (bits, elements, characters or blocks) to the total number transmitted.

Facsimile (also called FAX)

The transmission of photographs, maps, diagrams and other graphic data by communications channels. The image is scanned at the transmitting site, transmitted as a series of impulses and reconstructed at the receiving station to be duplicated on paper.

Feedback

The return of part of the output of a machine, process or system to the input, especially for self-correcting or control purposes.

Fiber

A single, separate optical transmission element, characterized by a core and a cladding.

Fiber Optics (Lightwave)

Light transmission through optical fibers for communication or signaling.

Four-Wire Circuit

A circuit containing two pairs of conductors, one pair for the transmit channel and the other for the receive channel. A communication path in which there are two wires for each direction of transmission.

Frame

A complete video image, consisting of two fields. Each NTSC frame is made up of 525 scan lines, half of which are allocated to each field. For full motion video, frames are transmitted at the rate of 30 per second. The European PAL standard dictates frames of 625 scan lines sent at the rate of 25 per second.

Frequency

The number of cycles per unit of time, denoted by Hertz (Hz).

Frequency Division Multiplex

A system of transmission in which the available frequency transmission range is divided into narrower bands, so that separate messages may be transmitted simultaneously on a single circuit.

Frequency Modulation (FM)

A method of modulation in which the frequency of the carrier is varied according to the amplitude of the transmitted signal.

FSK

Frequency **S**hift **K**eying, the most common form of frequency modulation, in which the two possible states (1/0, on/off, yes/no) are transmitted as two separate frequencies.

Full Duplex

Used to describe a communications system or component capable of transmitting and receiving data simultaneously.

Gigabits per Second (Gb/s)

One billion bits of digital information transmitted per second.

Gigahertz (GHz)
A unit of frequency equal to one billion Hertz.

Half-Duplex
Used to describe a communications system or component capable of transmitting data alternately, but not simultaneously, in two directions.

Hertz (Hz)
Synonymous with cycles per second: a unit of frequency, one Hertz is equal to one cycle per second.

High Definition Television (HDTV)
A broadcast television system that calls for the transmission of 1,125 line frames, developed by NHK, the Japanese broadcasting company.

Hybrid
A bridge-type device used to connect a 4-wire line to a 2-wire line so that both directions of transmission on the 4-wire line are isolated from each other, but are connected to the 2-wire line.

IEEE
The Institute of Electrical and Electronics Engineers, a publishing and standards-setting body responsible for many standards in the communication industry.

Inductance
A measure of the ability of a coil of wire, called an inductor, to block high-frequency signals from flowing through it. Inductance is measured in units of Henrys (H).

Information Bit
A bit used as part of a data character within a code group (as opposed to a framing bit).

Infrared
The band of electromagnetic wavelengths between 0.75 micron and 1,000 microns.

Intelligent Terminal

A "programmable" terminal which is capable of interacting with the central site computer and performing limited processing functions at the remote site.

Interface

A shared connection or boundary between two devices or systems. The point at which two devices or systems are linked. Common interface standards include EIA Standard RS-232B/C, adopted by the Electronic Industries Association to ensure uniformity among most manufacturers.

International Telecommunications Union (ITU)

An agency of the United Nations, headquartered in Geneva, Switzerland, to carry out studies of world telecommunications and make recommendations for standardization.

Kilohertz (kHz)

A unit of frequency equal to 1,000 Hertz.

Kilometer (km)

1,000 meters or 3,281 feet (0.621 mile)

Laser

An acronym for **L**ight **A**mplification by **S**timulated **E**mission of **R**adiation. A source of light with a narrow beam and a narrow spectral bandwidth.

Leased Channel

A point-to-point channel reserved for the sole use of a single leasing customer. See dedicated link.

Light Emitting Diode (LED)

A semiconductor device that emits incoherent light.

Lightwave (see Fiber Optics)

Loopback Tests

A test procedure in which signals are looped from a test center through a modem or loopback switch and back to the test center for measurement.

Megabits per Second (Mb/s)

One million bits of digital information transmitted per second.

Megahertz (MHz)

Unit of frequency equal to one million Hertz.

Metropolitan Area Network (MAN)

A group of local area networks (LANs) connected together over a distance of up to 50 kilometers.

Micron (μm)

Micrometer. Millionth of a meter = 10^{-6} meter.

Microsecond (μs)

One millionth of a second = 10^{-6} second.

Microwave

Any electromagnetic wave in the radio-frequency spectrum above 890 Megahertz.

Milliamperes (mA)

The measure of the current flowing in an electrical circuit. One thousandth of an amp. 1,000 mA = 1 A

Millisecond (ms)

One thousandth of a second.

Mode

A method of operation (as in binary mode).

Modem

A contraction of **mo**dulator / **dem**odulator.

Modulation

The process whereby a carrier wave is varied as a function of the instantaneous value of the modulating wave.

Multimode Fiber

An optical waveguide which allows more than one mode to propagate. Either step-index or graded-index fibers may be multimode.

Multiplex

To interleave or simultaneously transmit two or more messages on a single channel.

Multiplexing

The process of dividing a transmission facility into two or more channels.

Multipoint Circuit

A circuit that interconnects three or more stations.

Nanometer (nm)

One billionth of a meter = 10^{-9} meter.

Nanosecond (ns)

One billionth of a second = 10^{-9} second.

Network

A series of points interconnected by communication channels. The switched telephone network consists of public telephone lines normally used for dialed telephone calls; a private network is a configuration of communication channels reserved for the use of a sole customer.

Network Management System

A comprehensive system of equipment used in monitoring, controlling and managing a data communications network. Usually consists of testing devices, CRT displays and printers, patch panels and circuitry for diagnostics and reconfiguration of channels, generally housed together in an operator console unit.

Noise

Generally, any disturbance that tends to interfere with the normal operation of a communication device or system. Random electrical signals, introduced by circuit components or natural disturbances, which degrade the performance of a communications channel.

NTSC

The North American standard for color television systems. Named after the **N**ational **T**elevision **S**tandards **C**ommittee. Calls for 525 line frames transmitted at the rate of 30 per second.

Octet (see Byte)

On-line System

A system in which the data to be input enters the computer directly from the point of origin (which may be remote from the central site) or the output data is transmitted directly to the location where it is to be used.

Ohm

The unit of measurement for the resistance (DC) and impedance (AC) of an electrical circuit.

Packet

A group of bits, including address, data and control elements that are transmitted and switched together.

Packet Switching

A data transmission method, using packets, whereby a channel is occupied only for the duration of transmission of the packet.

Parity Check

A checking system that tests to ensure that the number of ones or zeros in any array of binary digits is consistently odd or even. Parity checking detects characters, blocks or any other bit grouping that contain single errors.

Phase Alternating Line (PAL)
The broadcast television standard in Europe calling for 625 lines per frame transmitted at 25 frames per second.

Photonics
A technology based on interactions between electrons and photons.

PIN Diode
A device used to convert optical signals to electrical signals in a receiver.

Polling
A centrally controlled method of calling a number of terminals to permit them to transmit information. As an alternative to contention, polling ensures that no single terminal is kept waiting for as long a time as it might under a contention network.

Pulse Code Modulation (PCM)
A modulating analog signal is sampled, quantized and coded so that each element of the information consists of different kinds or numbers of pulses and spaces.

Real Time
Generally, an operating mode under which receiving the data, processing it and returning the results takes place so quickly as to actually affect the functioning of the environment, guide the physical processes in question or interact instantaneously with the human user(s). Examples include a process control system in manufacturing, or a computer-assisted instruction system in an education institution.

Refractive Index
The ratio of light velocity in a vacuum to its velocity in the transmitting medium.

Repeater

A device in which signals received over one circuit are automatically repeated in another circuit or circuits, generally amplified, restored or reshaped to compensate for distortion or attenuation.

Response Time

The time a system takes to react to a given act; the interval between completion of an input message and receipt of an output response. In data communications, response time includes transmission times to the computer, processing time at the computer (including access of file records), and transmission time back to the terminal.

Serial Transmission

A mode of transmission in which each bit of a character is sent sequentially on a single circuit or channel, rather than simultaneously as in parallel transmission.

Simplex

Generally a communications system or device capable of transmission in one direction only.

Singlemode Fiber

A fiber waveguide on which only one mode will propagate, providing the ultimate in bandwidth. It must be used with laser light sources.

Spectral Bandwidth

The difference between wavelengths at which the radiant intensity of illumination is half its peak intensity.

Synchronous Transmission

A transmission method in which the synchronizing of characters is controlled by timing signals generated at the sending and receiving stations (as opposed to start / stop communications). Both stations operate continuously at the same frequency and are maintained in a desired phase relationship. Any of several data codes may be used for the transmission, so long as the code utilizes the required line control characters (also called bi-sync or binary synchronous).

Switched Telephone Network

A network of telephone lines normally used for dialed telephone calls. Generally synonymous with the direct distance dialing network, or any switching arrangement that does not require operator intervention.

Terminal

Any device capable of sending and/or receiving information over a communication channel, including input to and output from the system of which it is a part. Also, any point at which information enters or leaves a communication network.

Time Division Multiple Access (TDMA)

A digital cellular standard. This technique is also used in satellite communications and allows more than one earth station access to a single satellite channel.

Time Division Multiplexer

A device which permits the simultaneous transmission of many independent channels into a single high-speed data stream by dividing the signal into successive alternate bits.

Trunk

A single circuit between two points, both of which are switching centers and / or individual distribution points.

Voice Grade Channel

A channel suitable for the transmission of speech, digital or analog data or facsimile, generally having a frequency range of about 300 to 3,000 Hertz.

Wavelength Division Multiplexing (WDM)

A technique which employs more than one light source and detector operating at different wavelengths and simultaneously transmits optical signals through the same fiber while message integrity of each signal is preserved.

Wide Area Network (WAN)

A telecommunications network that covers a large geographic area. It typically links cities, and may be owned by a private corporation or by a public telecom operator.

Wideband Channel

A channel broader in bandwidth than a voice-grade channel (see Broadband Channel).

X.25

ITU's international standard which defines the interfaces between a packet-mode user device and a public data network.

X.75

ITU's international standard for connecting packet switched networks.

Bibliography

Brooks, John, Telephone: The First Hundred Years,
New York: Harper and Row, 1976.

Martin, James, Telecommunications and the Computer
(second edition), Englewood Cliffs, NJ: Prentice-Hall,
1976.

McDermott, T. C., Gigabit Lightwave Technology,
(523-0610029-001K3J) 1985, Alcatel Network Systems,
Richardson, Texas.

McGovern, Tom, Data Communications,
Scarborough, Ontario: Prentice-Hall Canada, 1988.

Nellist, John G., Fiber Optic System Availability,
Fibresat 86, Expo 86, Vancouver, Canada, September
9 – 12, 1986.

Ponder, D. E., Current Issues in High Capacity Lightwave
Transmission Systems Implementation,
(523-0610046-001K3J) 1985, Alcatel Network Systems,
Richardson, Texas.

Reeve, Whitham D., Subscriber Loop Signaling and
Transmission Handbook – Analog,
Piscataway, NJ: IEEE Press, 1992

Reeve, Whitham D., Subscriber Loop Signaling and
Transmission Handbook – Digital,
Piscataway, NJ: IEEE Press, 1995

Talley, David, Basic Telephone Switching Systems,
Rochelle Park, NJ: Hayden, 1979.

Tomasi, Wayne, Telecommunications,
Englewood Cliffs, NJ: Prentice-Hall, 1988.

Index

John (Jack) Nellist

As General Transmission Engineering Manager at the British Columbia Telephone Company, Jack led the development and implementation of fiber optic technology into the British Columbia Telephone Network.

He was appointed overall coordinator for the first field trial of a fiber optic transmission system in 1978. Jack presented one of the first papers on an operating fiber optic system in 1979 at the International Fiber Optic Conference held in Chicago, Illinois.

He received the "Outstanding Technical Achievement Award" in April 1980 by the Applied Science Technologists of British Columbia, in recognition of technical competence in the field of fiber optics.

As a communications consultant, he participated in the marketing, design and installation of many fiber optic projects provided by Fluor Daniel Telecommunications Services Division of Irvine, California, from 1984 to 1986.

With more than 40 years of telecommunication experience, Jack has published and presented numerous technical papers and articles on various telecommunication subjects.

Answers to Review Questions

Section 2

1. 30 Hz to 16,500 Hz

2. 200 Hz to 5,000 Hz

3. 300 Hz to 3,400 Hz

4. Power is provided by a 48 volt battery and fed to the telephone over a twisted pair of copper wires.

5. Electrets

6. Dial pulses and tones

Section 3

1. Codec

2. No. Music, video or photographs can be digitized.

3. 4kHz

4. 64 kb/s

5. A bit is always one of two things. Pulse or no pulse, mark or space, 1 or 0, on or off.

6. The degradation of the received signal does not alter the information content until the receiving equipment reads a pulse as no pulse, or vice versa.

7. The repeater reads its input signal, extracts the information, and uses it to generate a brand new output signal.

Section 4

1. FDM and TDM

2. To permit a number of voice channels to be transmitted over the same line at the same time.

3. PCM combined with TDM

4. TDM – a piece of each channel in turn by time

5. Sampling, quantizing and encoding.

6. 8 kHz

7. 193

8. 6,000 feet

Section 5

1. End office, toll center, primary center, sectional center and regional center.

2. No. It can also home on a class 1, 2, or 3 office.

3. This is the gateway between the interexchange carrier and the exchange carrier's end office.

4. A high-usage trunk connects to the lowest terminating office in the hierarchy. If all high-usage trunks are busy, then the call ascends to the highest level office over a final trunk.

5. Dynamic nonhierarchial routing and dynamically controlled routing.

6. CCIS is a system for exchanging signaling information between exchanges via a network of signaling links instead of on the individual voice circuits.

Section 6

1. DS-1, 1.544 Mb/s
    DS-2, 6.312 Mb/s
    DS-3, 44.736 Mb/s

2. M-23 multiplexer takes one bit at a time from each of the seven DS-2 inputs and sequentially interleaves them to form a DS-3 output.

3. Twelve

4. DSI and LBRE

5. Adaptive differential pulse code modulation

Section 7

1. DDD is the term used to describe long distance calls dialed by customers without assistance by operaters.

2. −12dBm

3. 45 ms

4. Loss in the subscriber loop

5. The digital echo canceller uses digital signal processing techniques which generate a replica of the echo. The echo replica is continuously subtracted from the transmit signal at the receiving end, thereby selectively cancelling the echo.

6. 3,000 kilometers (1,850 miles)

Section 8

1. 8.0 db

2. 1,300 Ohms

3. 20 mA

4. 22 is the wire gauge, H indicates 6,000 feet spacing and 88 signifying the inductance.

5. End office

6. No

7. 128

8. 1.5 Mb/s

9. 12,000 feet

10. One, 18,000 feet

11. 2.4 to 64 kb/s

Section 9

1. Digital microwave systems detect and reshape the transmitted pulses at all repeater stations.

2. 1,344

3. 26 miles

4. Only one user can occupy a frequency band between the same locations. Multipath fading. Digital radio must co-exist with existing analog radio systems.

5. Parabolic dish antennas operate in one frequency band. Horn antennas permit operation in several frequency bands.

6. The right-of-way for a microwave system is free.

7. 44,000

8. 622 Mb/s

Section 10

1. A satellite whose orbit is synchronozed to the earth's rotation.

2. 320 kilometers (200 miles)

3. 35,880 kilometers (22,300 miles)

4. The transponder receives the signal from the earth station, amplifies the signal, changes the frequency and retransmits the signal back to earth.

5. One color TV signal and 1,200 voice circuits.

6. The size and shape of the satellite's radio beam on the surface of the earth.

7. 540 ms

8. They use smaller and less expensive antennas and the band does not coincide with those used for terrestrial microwave systems.

9. Very small aperture earth terminal

10. The uplink signal consists of a set of digital data bursts, each burst containing data addressed to a particular receiving earth station. The burst is assigned to a time slot, and the earth station clocks are synchronized, one time slot for each earth station in rotation.

11. Twenty-four

12. 20,200 kilometers

Section 11

1. Under direct control of the dial pulses from the calling telephone, the Strowger switch connects pairs of wires by step-by-step operation of several series switches operating in tandem.

2. Electrical and mechanical noise; high maintenance costs; call blockage.

3. 20 input lines and 10 output lines.

4. Each telephone call is assigned a separate physical path through the switch.

5. 128

6. 53,760

7. The remote portion of a digital switch located closer to the customer.

8. One terabit per second

Section 12

1. A key telephone

2. Voice signals are switched in their original analog form through an analog switch. Voice signals are first converted to digital form by a codec before being switched through a digital switch.

3. The central processing unit (CPU); the network; the peripheral unit.

4. Trunks connect the PBX to the local central office.

5. 60,000

Section 13

1. 10:00 AM to 11:00 AM

2. 10% to 12%

3. The number of calls times their average duration in seconds divided by 100.

4. 36 CCS

5. Five calls out of every 100 attempted will get blocked.

6. 150

7. Four

Section 14

1. A codec converts analog information into a digital format. A modem converts digital information into an analog format.

2. Yes

3. The USART converts the parallel transmission to a serial bit stream.

4. To check for errors.

5. In asynchronous transmission each character is transmitted independently of all others. Synchronous operation is used for the transmission of a block of characters.

6. A network, using packets, where a channel is occupied only for the transmission of the packet.

7. Datapac

8. 28.8 kb/s

9. 56 kb/s or 64 kb/s

10. X.25

11. With X.25, error checking is performed at every link, whereas in frame relay it is done only at each end.

12. Permanent virtual connection (PVC) and switched virtual connection (SVC).

13. 1,500 bytes

14. 1.5 Mb/s

Section 15

1.   59 Mb/s

2.   90 Mb/s

3.   Video codec

4.   The NTSC standard calls for 525 line frames transmitted at the rate of 30 per second. The PAL standard calls for 625 line frames transmitted at the rate of 25 per second.

5.   56 kb/s, 64 kb/s, 384 kb/s and 1,536 kb/s

6.   800 to 1

7.   Multimedia integrates the worlds of video transmission and digital communications with the world of computing.

8.   Joint Photographic Experts Group (JPEG) and Moving Pictures Experts Group (MPEG).

Section 16

1.   1 to 12 miles

2.   Growth is provided by cell splitting and adding more transmitters.

3.   Mobile telephone switching office (MTSO)

4.   800 to 900 MHz

5.   832

6.   The serving cell site examines the signal strength every few seconds and transfers the call to a cell site closer to the mobile unit if the signal level becomes too low.

7.   66

8.   A new generation of low-cost, compact digital cordless telephones for pedestrian and residential use.

9. (a) High density, on premises, low power systems
   (b) High speed, wide-coverage vehicular systems
   (c) High density, wide coverage cellular systems

10. The first multinational standard; cordless telephone 2 common air interface (CT2 CAI)

Section 17

1. The fiber, the buffer tube and the jacket.

2. Loose buffer and tight buffer.

3. Numerical aperture is a measure of a fiber's light acceptance angle.

4. The first window is at 850 nm, the second at 1300 nm and the third at 1550 nm .

5. Multimode dispersion is caused by some light rays that travel straight down the core and others, following a zigzag path, are therefore somewhat delayed.

6. Multimode step index, multimode graded index and singlemode step index.

7. The zero dispersion point has been shifted from 1313 nm to 1550 nm.

8. The SMA, the FC, the ST, the SC and the biconic.

9. Light emitting diode (LED) and the laser diode.

10. PIN diode and the avalanche photo diode (APD).

11. The FP laser has a spectral width of several nanometers whereas the DFB laser is similar to a single frequency source with a spectral width of 0.5 nm.

12. Lightwave frequencies extend from 10^{14} to 10^{15} Hz. This is beyond the sensitivity of the human eye.

13. 8,064

14. Five errors per 10 billion bits

15. To determine the performance parameters of the system and the available operating margin.

16. WDM employs more than one wavelength over the same fiber while preserving message integrity of each signal.

17. A pump laser operating at 1480 nm induces gain in a fiber doped with a small amount of erbium.

18. 24

19. 51.8 Mb/s

20. OC-48 operating at 2488.32 Mb/s.

21. One protection line for N regular working lines.

22. Placing three or four one inch plastic tubes in a standard four inch duct.

23. 1.2 meters (4 feet)

24. In the fusion technique, the fiber ends are heated and fused together by an electric arc.

25. Epoxy and polish splices, cleaved splices using ultraviolet curable epoxies, and mechanical cleave and crimp splices.

26. An OTDR launches narrow laser pulses into a fiber and monitors the reflected light with respect to time.

27. Fresnel reflection occurs as light passes into a medium having a different index of refraction. Rayleigh scattering is caused by light striking glass molecules and being scattered in all directions.

28. Passive Optical Network (PON)

29. The ONU receives and transmits a packetized optical signal. It performs the electro-optical and digital to analog conversions and provides the interface between the fiber system and the customers' copper drop.

Section 18

1. 2 km

2. 10 Mb/s

3. Coaxial cable and unshielded twisted pair wiring (UTP)

4. Both

5. Bus

6. It can survive a cable cut and still provide service to all nodes.

7. Fiber distributed data interface

8. 100 Mb/s

9. A riser cable goes from the building entrance splice to the user's desk top equipment.

10. Interconnect cables are used to link optical equipment.

11. 900 micron PVC

12. Two B channels will carry 64 kb/s transmissions of voice, data, etc. and one D channel will carry 16 kb/s of control information.

13. 1.536 Mb/s

Section 19

1. Interactive and distributive

2. 53

3. Virtual channel identifier (VCI) and virtual path identifier (VPI)

4. The network node interface is the interface between the transmission facility and the network node.

5.   The user network interface is the interface between the user equipment and the network termination.

6.   Switched multimegabit data service

7.   9188

8.   DS-1 and DS-3

9.   An ATM switch uses the VCI and VPI. An ATM cross connect uses only the VPI.

10. Video

11.   155 Mb/s (OC-3), 622 Mb/s (OC-12) and 2.4 Gb/s (OC-48).